Policies lost in translation?

Unravelling water reform processes in African waterscapes

Policies lost in translation?

Unravelling water reform processes in African waterscapes

DISSERTATION

Submitted in fulfillment of the requirements of
the Board for Doctorates of Delft University of Technology
and of the Academic Board of the UNESCO-IHE
Institute for Water Education
for the Degree of DOCTOR
to be defended in public on
Thursday, 10 December 2015, 15:00 hours
in Delft, the Netherlands

by

Jeltsje Sanne Kemerink-Seyoum

Master of Science in Civil Engineering, Delft University of Technology
born in Krimpen aan den IJssel, the Netherlands

This dissertation has been approved by the promotors:

Prof. dr. ir. P. van der Zaag and dr. ir. R. Ahlers

Composition of Doctoral Committee:

Chairman	Rector Magnificus TU Delft
Vice-Chairman	Rector UNESCO-IHE
Prof. dr. ir. P. van der Zaag	TU Delft / UNESCO-IHE, promotor
Dr. ir. R. Ahlers	Independent Researcher, co-promotor

Independent members:

Prof. dr. E. Manzungu	University of Zimbabwe, Harare, Zimbabwe
Prof. dr. F.D. Cleaver	University of Sheffield, Sheffield, UK
Prof. dr. ir. W.A.H. Thissen	TU Delft
Prof. dr. ir. M.Z. Zwarteveen	UvA / UNESCO-IHE
Prof. dr. ir. H.H.G. Savenije	TU Delft, reserve member

CRC Press/Balkema is an imprint of the Taylor & Francis Group, an informa business

© 2015, J.S. Kemerink-Seyoum
Painting cover by Anne Kemerink

Published by:
CRC Press/Balkema
PO Box 11320, 2301 EH Leiden, the Netherlands
e-mail: Pub.NL@taylorandfrancis.com
www.crcpress.com – www.taylorandfrancis.com

ISBN 978-1-138-02943-9 (Taylor & Francis Group)

To Helina and Nahom,

may you strive for equity, in whatever form, in whatever way

Abstract

Since the 1980s a major change took place in public policies for water resources management. The role of governments shifted under this reform process from directing, and investing in, the development, operation and maintenance of water infrastructure to managing water resources systems by stipulating general frameworks and defining key principles for water allocation. This change in policy approach has been criticized based on empirical research which shows that policies often do not achieve what they envision on paper due to interpretation, negotiation and rearrangement by socially positioned actors at different spatial levels leading to uncertain, hybrid and context specific outcomes. However, at the same time, empirical research shows that the new policy approach has paved the way for the proliferation and implementation of similar policy models in dissimilar contexts that reinforces inequities in terms of access to and control over water resources in waterscapes around the world. This dissertation aims to unravel this scientific paradox by studying to what extent, how and why the mainstream approach in water policy reforms influence, shape and change the water resource configurations within waterscapes. To do so, this research examines the interplay between public policies designed and implemented by government agencies and the institutions that govern access to and control over water resources among groups of agricultural water users. How this relationship unfolds within waterscapes that are historically constituted by natural and social processes is the overarching research objective of this interdisciplinary study. For this purpose, this research analyzes case studies in four African countries that have reformed their water policies during the last decades, namely Kenya, South Africa, Tanzania and Zimbabwe. The water reforms in these countries have all been instigated by the global shift in public policy approach and as such share similar narratives to justify the reform processes as well as aim to achieve similar objectives.

This research builds further on critical institutionalism (e.g. Cleaver, 2002; 2012; Cleaver and De Koning, 2015), a school of thought that conceptualizes institutions as outcomes of dynamic social processes that shape, regulate and reproduce human behaviour across time and space. This theory helps to explain why and how processes of institutional change produce different outcomes for diverse social groupings in society. To understand the contemporary policy making processes this research adopts a political perspective in which policies are seen as outcomes of a discursive practice of policy networks that frame problems and ideas, construct policy narratives and disseminates policy models (e.g. Conca, 2006; Rap, 2006; Molle, 2008; Peck and Theodore, 2010). Moreover, this research employs the concept of waterscape, in which social relations and natural processes are understood to concurrently constitute and reorder physical environments (e.g. Swyngedouw, 1999; Budds, 2008; Mosse, 2008). This concept is useful to analyze how the outcome of the interplay between existing institutions and policy interventions materializes within historically produced landscapes and is affected by physical artefacts such as hydraulic infrastructures. This research uses the extended case study method (e.g. Burawoy, 1991; 1998) to analyze the water reform processes in catchments located in the four African countries. The research data is primarily obtained through 175 semi-structured interviews with large-scale and small-scale farmers and other actors located within the catchments, which is complemented with data obtained through focus group discussions, informal conversations, field observations and attendance of meetings as well as analyzing policy documents, maps, satellite images, databases, scientific publications and project reports.

Each of the four case focuses on different facets of the reform process in order to thoroughly comprehend the working and implications of the shift in the policy approach that took place since the 1980s. The Tanzanian case study focuses on the negotiations over access to water between and within traditional smallholder irrigation systems during the onset of the water reform process. This case shows the hybrid and dynamic nature of institutions that govern water resources as well as how these institutions endure and evolve over time. It gives a detailed account of how water users use different normative frames from various sources to legitimize their claims in negotiations over access to and control over water and how this plural legal reality affects sharing of water between the farmers. The case study located in South Africa illuminates the contested nature of water reform processes and how this shapes the interactions between large-scale and small-scale farmers. This case study shows that the internationally praised South African Water Act is based on different, partly conflictive, normative understanding and discusses how this leads to a partial implementation within the still highly segregated South African society. Moreover, this part of the research analyzes how the use of seemingly neutral policy models, in this case the decentralization through establishment of water users associations, leads to the reinforcement of structural inequities in terms of access to and control over water resources in the case study catchment. The case study in Kenya focused on the rationales used to justify the water reform process and unravels to what extent these rationales are valid for various kinds of water users in the case study catchment. It shows that only a few historically advantaged commercially oriented farmers benefited from the new legislation in the case study catchment, either by adapting to or by rejecting the water reform process. In particular, this case study identifies several unexpected and undesired outcomes of the reform process for small-scale farmers who are member of water user associations and shows how this is linked with the institutional plurality as well as the type of hydraulic infrastructure these farmers have access to. The last case located in Zimbabwe studies the implications of the implementation of water reform policies in a catchment within a rapidly changing context due to instability in land tenure and collapse of the national economy. This case study show how people respond to the changing conditions, including the water reform process, by reordering their physical environments and moving their agricultural activities upstream, where their water use is regarded illegal yet cheaper and more secured. Moreover, this case study explores the use of satellite images to incorporate complex socio-nature processes into policy making process to aid policy makers who wish to respond to dynamic and context specific circumstances.

In the final chapter of this dissertation the extended case studies brought together in an incorporated comparison (McMichael, 1990; 2000) based on the shared epistemic origin of the reform processes which has led to similarities in the narratives to justify the reforms as well as the objectives set and means selected for the reform process. Based on this comparison this dissertation shows that the water reform processes in the case study countries contribute to processes of social differentiation that mainly benefits historically advantaged individually operating water users who produce their crops for the commercial market. This research shows that the institutions governing the water resource configurations in the case study areas are dynamic in nature, constantly negotiated, reconfirmed and contested within the interactions among the farmers. In this process farmers actively use the normative frames and institutional blueprints that have been introduced by the national governments as part of the water reform process. They, consciously and unconsciously, have interpreted, reworked, adopted and rejected parts of the government's policies and combined them with existing institutions into new hybrid institutions. Also government officials actively participate in this process, trying to manipulate and craft institutions in an attempt to not only pursue the stated and unstated policy objectives but also to suit their own understandings and interests. Once enacted, policies thus add to the legal repertoire actors can draw on in a continuous bargaining process to establish the institutions that determine access to, control over and distribution of water resources. Since the agency of actors is neither rigid nor equal, disparately circumscribing their capability to respond to and manipulate policy interventions, this dissertation concludes that the water reform policies have altered the institutions that govern water resource configurations through uneven processes of bricolage (e.g. Cleaver, 2002; 2012).

The findings of this research show that policies do have agency within waterscapes, especially when they are aligned with the interests of the elite and rolled out through seemingly neutral or even 'progressive' policy models. As such policies can only to a limited extent contribute to progressive societal change, especially in this neoliberal era as the interests of influential actors operating within national and international policy arenas are so tied up and fixed within a particular normative understanding of the world. This dissertation shows the implications of this neoliberal inclined shift in public policies towards primarily attempting to steer institutional processes and excluding technological policy instruments such as investments in the development of hydraulic infrastructure. Since the majority of the agricultural water users in the case study countries lack access to (adequate) hydraulic infrastructure, large parts of the water legislation enacted under the reform processes is not attainable for them and, in some cases, even lead to undesired outcomes such as further marginalization and transformation of the waterscapes. This shows a disjuncture between the policy objectives and the selected instruments to achieve these objectives. Moreover, this research shows that excluding targeted investments in the development of hydraulic infrastructure for historically disadvantaged groups has severely narrowed the options and thus the capacity of the governments to redress the colonial legacy since without these investments the small-scale farmers have little chance to increase their water use and move their livelihood beyond subsistence.

This dissertation contributes to existing theories and concepts related to institutional processes and water governance, and particularly to advance critical institutionalism. This research contributes to enrich this theory in four ways, namely by including the implications of structural configurations of institutional processes at larger spatial scales on how water reform processes unfold within contextualized waterscapes. This is achieved by selecting extended case studies and incorporated comparison as methodological approaches that help to understand the interactions between processes at various spatial levels as well as by linking critical institutionalism to theories that explain the political nature of contemporary policy making processes. Second, this dissertation advances critical institutionalism by adopting a socio-nature perspective and specifically looking at how the physical environment constitutes social relations. The concept of waterscapes is used to include the agency of hydraulic infrastructure as well as the materiality of water in shaping institutions that govern water resource configurations within waterscapes. Third, this dissertation analyzes the normative perspectives underlying policy interventions in relation to the normative orders that prevail is society. In this way it captures not only how authority is possessed and exercised by actors, but also role of norms and institutions in this process in producing, maintaining and contesting structural inequities in society (e.g. Foucault, 1979, 1980). And last, this research contributes to critical institutionalism by attempting to show how the findings of these kinds of studies can be useful for policy makers. For this purpose this dissertation includes concrete suggestions for revisiting the current water policies in the case study countries, namely by embracing the political nature of the policy making process through critical policy analysis; by engaging in a more profound implementation and learning process to assess contextual implications of and responses to reform processes; and by adopting a comprehensive policy approach that includes institutional, financial and technological policy instruments.

Based on this research recommendation are made for further research, including ethnographic research on the actors involved in the policy networks that disseminate the mainstream policy models as well as research on the agency of the physical environment on shaping social relations. This dissertation ends with a critical reflection on the research by discussing how theoretical and methodological choices made within this research shaped the findings of this study.

Acknowledgement

Perhaps the most important section of a dissertation, the acknowledgement; after all it is the last part to write for most PhD candidates and the first words to read by many. Therefore I better do my utmost best to make this section into an interesting and inclusive read.

Even though some say conducting PhD research can be a lonely journey, I never experienced it as such. This is largely because this dissertation is the result of a collaborative effort of many, in which I was perhaps the spider in the web but for sure not the sole researcher. As such, I interacted with supervisors, students, co-researcher, peer-reviewers and journal editors. Moreover, since in this research I adopted a social science approach that encourages active engagement between the researcher and the subject of the study, I had the pleasure to have many cherished encounters and lively discussions with the people that shaped the processes I studied, in particular the farmers in the catchments where this research is situated. The part time basis on which I carried out this research also meant that I continued to be a colleague, a lecturer, a programme coordinator, an employee, a project team member and a mentor, roles that kept me far from isolation. Lastly, of course I maintained a life beyond my career in which I was, and luckily still am, a daughter, a sister, an aunt, a friend, an occasional football or volleyball player, a cousin, a travel companion, a neighbour, and recently became a mother, a wife and a daughter-in-law. These social identities and associated relationships kept me afloat when I needed it most. In other words, I have many people to thank who supported me on this eight year long journey.

First of all I would like to thank UNESCO-IHE for offering the opportunity to conduct this PhD research while being employed as a lecturer by funding part of my research activities and allowing me to write some 'unbillable' time while finalizing this dissertation. For the case studies carried out in Tanzania and South Africa I would like to thank the funding agency of the SSI programme, the Department of International Cooperation of the Netherlands (DGIS). The research in South Africa was also partly funded by the Agris Mundus programme of the European Commission.

I am greatly indebted to my promotor Professor Pieter van der Zaag for his enthusiasm and continuous support on this long journey. Probably I was not one of your easiest students to guide, but the freedom you gave me to explore and discover my research interests, yet at the same time constantly bringing my feet down into the muddy physical reality of the irrigation systems, was exactly what I needed to complete this study. I admire your vast knowledge, your quick understanding and your devotion to the African continent. Your input, as well as your friendship, has been essential for both the content of this research as well as my personal growth. I also thank Marlou for the pleasant conversations, the delicious meals and driving Pieter and me to the airport and back at impossible times.

I am extremely thankful to my co-promotor Rhodante Ahlers for letting me borrow her critical mind. Even though I take full responsibility for the content of this dissertation, without your sharp analytical skills this research would not nearly have been what it is now. I greatly appreciate your passion for research and the ways in which you strive for equity. Sometimes I felt overwhelmed by your feedback on my work; it took a few days to get myself together again and to begin to grasp what you were trying to explain to me, but I needed that push to move from being an engineer into being a crossbreed of different disciplines. Besides being an admirable researcher, I got to know you as a very affectionate person with a good sense of humour, who did not only listen to my struggles at work but also the challenges I faced in my personal life.

I also would like to thank the Master students who directly or indirectly contributed to the implementation of this research, in particular Lukas Kwezi, Linda Méndez, Ndakaiteyi Chinguno, Abeer Al-Asady, Stephen Ngao Munyao and Abeer Mahmoud. Without you this research would not have been feasible since you have been, at least partly, my eyes and my ears in the field. I very much enjoyed our interactions, I remember many lively discussions on how we could pose a question and how we should interpreted the answer. This research has truly been a joint learning process. I am still grieving the sudden death of Ndaka in August this year at the early age of 41, leaving behind her beloved husband and three children. I will greatly miss her bright mind, warm personality and good sense of humor. I would like to acknowledge the contributions of the co-authors of the articles published under this research and express my gratitude to the (anonymous) external reviewers and journal editors who gave constructive comments on earlier versions of the individual chapters within this dissertation.

Above all, I would like to express my sincere appreciation for the people residing in the case study areas that have been willing to free up time to share their experiences and opinions with me and my co-researchers. The conversations we had, while sitting under mango trees, walking along the irrigation furrows or harvesting maize on your fields, belong to the most preciously remembered memories in my life, I thank you for that. I am also thankful to the other respondents, including government officials, NGO staff, extension officers and fellow scholars, who have been very open and outspoken during the interviews conducted for this research.

I would like to express my gratitude to the SSI research team for adopting me as one of them even though my research only fitted partly under their programme objectives. For the fieldwork conducted in Tanzania I would like to particularly thank Marloes Mul, Elin Enfors, Hodson Makuria and Hans Komakech for their crucial guidance in starting up this research, for sharing their vast knowledge on the catchment and for their good company in the field. To be more specific on that last aspect: Marloes, thanks for teaching all the kids in the catchment to say *doei*; Elin, thanks for serving delicious breakfast with Swedish imported *knäckebröd* in your beautiful home in Bangalala village; Hans, thanks for being my indispensable bodyguard in the local nightclub; and *Mzee*, thanks for occupying the small village fridge with your two daily *Serengeties* (and thanks for smuggling *samaki kidogo sana* for me into the Netherlands). Implementation of the fieldwork in Tanzania was assisted by the Soil-Water Management Research Group of Sokoine University of Agriculture. My gratitude goes to Patricia Kadeghe and Gevaronge Myombe for translating the interviews and group discussions as well as arranging the logistics in the field. For the fieldwork in South Africa I am indebted to Professor Graham Jewitt, Michael Malinga, Victor Kongo, Job Kosgei, Maxwell Mudhara and Rebecca Malinga for introducing me to the farmers, facilitating my research activities and making me feel at home in Pietermaritzburg. The implementation of the fieldwork was assisted by the School of Bioresources Engineering and Environmental Hydrology and the Centre for Environment, Agriculture and Development of the University of KwaZulu-Natal. My gratitude goes to Hlengiwe Mabaso for translation during the interviews and group discussions. The fieldwork in Kenya was assisted by the Water Resource Management Agency through the Ewaso Ngíro North Catchment Area Office and special gratitude goes to the chairperson of Likii RWUA for facilitating the research process. And for the fieldwork conducted in Zimbabwe I would like to thank the Save Catchment Council, the Odzi

Acknowledgement

Subcatchemnt Council, the ZINWA Mutare Hydrology Section, the Irrigation Management Commitee of the Nyanyadzi scheme and the Agricultural Extension Office.

There are many colleagues at UNESCO-IHE to thank for their support during this PhD trajectory; either for listening to my occasional frustrations or by being patient and not overloading me with additional tasks or for accepting my unsocial behaviour such as having lunch behind my desk. I would like to mention a few in particular. First of all Erwin, my first boss at UNESCO-IHE, who always showed his confidence in me and who encouraged me to pursue an academic career even if that meant I would have to leave his group. You are still somebody I rely on when I seek advice on something at work and I am glad we still catch up over lunch once in a while. I am sure my family is also grateful to you for making the Dutch summary of this dissertation somehow understandable. Secondly I am greatly indebted to Klaas for being my big brother and my 'institutionally fuzzy' mentor in many aspects of life. I guess no longer sharing an always lively and amazingly messy office with you has helped me in making progress with this dissertation, though I have fond memories of being roommates including making 'print screens' on which the top of your head is still visible. Erwin and Klaas, I am honoured that the two of you have accepted to be my *paranymphs* during the doctoral defence ceremony, having you stand in front of me during these sixty anxious minutes will make me hopefully feel at least a bit safe. In addition I like to thank my other bosses, Margreet, Pieter, Frank and Jan, for the opportunities they have given me and the support they provided to conduct this research. Margreet, I am very happy you made the brave decision to head our somewhat eccentric water governance chair group. I would like to thank Michelle and Susan for sharing so much more than just work, Hermen for being not only a beloved colleague but also for being my PhD comrade, Maria for her creativity and unconventional mentality and Anne for being such a kind and perceptive person. I am grateful to many other colleagues, amongst others Wim, Jetze, Mireia, Vanessa, Mishka, Zaki, Jaap, Chris, Berthold, Yasir, Schalk-Jan, Tineke, Berry, Martin, Ilyas, Erik, Ioana, Micha, Edwin, Jochen, Annelieke, Sylvia, Charlotte, Raquel, Maria, Gretchen, Caroline, Marleen and Robert, for making my working days so much more pleasant. I also would like to thank colleagues outside UNESCO-IHE for being a source of inspiration, in particular Alex, Frances, Edwin and the members of the WaterNet network in Southern Africa, including Themba, Jean-Marie, Dominic, Dinis and my late hero Lewis. I acknowledge the support of the UNESCO-IHE library staff for helping me to get access to even the weirdest publications, I thank Anique for guiding me through the ever changing administrative procedures that PhD candidates need to follow and Peter for lay-out of this dissertation. I am also grateful to the various batches of Water Management students I had the privilege to teach as preparing the lectures for you as well as the discussion we had in class helped me in constructing the argument for this dissertation.

With my friends I did not talk frequently about my PhD research, simply because we had too many other things going on in our lives to talk about. Nevertheless, or perhaps therefore, they have been very important for me to keep my sanity during the past years. I would like to thank Brenda and Betty in particular for being there for me when I needed a shoulder to lean on and daring to be critical when I made mistakes in life. I also like to thank Renske for being such an amazing listener. I always feel good after the little time we get together, you are like family to me. I would like to make use of this opportunity to express my gratitude to my friends in South Africa, in particular Msa, Thecla and Mike, who opened their homes to me, provided me with insights into the dynamic social context in which my research is situated and made my stay in Johannesburg so much more fun. I am grateful to Tinie, Remigio and Ivo for welcoming me to their home in Maputo and visiting me in South Africa when I was in need. I thank my friend Lindsay for staying close even though she moved far away and I would also like to thank my other friends, including Milli, Mahlet, Pato, Aki, Angela, Gaetano, Michelle, Liselotte, Lotte, Kristina, Judit, Arlex, Nathasja, Seleshi, Ledetta, Kees and Rosa, for their enduring friendship. Hopefully now that finally this 'baby' is delivered we can have some more time to together.

Where to start when expressing my gratitude to my parents, Peter and Anne, for all that they have done for me and for all that they mean to me? Thanks to you I have very dear memories of my youth, playing outside all day long building castles with water and sand or making huts high up in the trees, and when I got tired and dirty I could just simply go home where it was always warm and cosy. Yet at the same time you made me conscious at an early age that the security I experienced was an exception and not a rule for kids growing up in this world. When I was just a few years old you took me along on a protest against nuclear weapons and world politics remained a frequent and hot topic of debate at our dinner table throughout my adolescent years. I am extremely grateful that you simulated my political awareness, curiosity and independent thinking, for sure that has shaped the focus of this dissertation. You are a role model for the kind of parent I hope to be and I love you with all my heart. If there is one person who taught me the essential skills in life it has been my sister Marijntje, from how to walk to how to distinguish between the colours red and blue, from how to jump over creeks with and without wetting my clothes to how to close the lid of the toothpaste tube, and from how to bake bread with coffee flavour to how to play with other kids; all competences proven to be essential for surviving in rural Africa. You have always been very close to me and I could not have wished for a nicer and more compassionate sister than you. Thank you, as well as my dear big brother Bert and my lovely nieces Pleuntje, Djouke, Hanne and Otje, for your unconditional love and support during the past years.

I would like to thank my extended family, my late grandparents, uncles, aunts and cousins for providing a solid and warm base to rely on. We do not see each other frequent, but the small gestures of compassion that come with the ups and downs in life go a long way and the annual camping holidays with some of you are always good fun. In particular I am grateful to my uncle Lambert for planting an African seed in my heart when I was a kid by telling me stories about his life in Mozambique. I would like to express my gratitude to my parents-in-law Dagnachew Seyoum and Askal Moges for welcoming a strange bird, who only speaks a few words Amharic and who eats *injera* in a funny way, into the family. Also my brothers-in-law, Dave, Mini, Aschu and Tedu and their families, have been of great moral support at different occasions during the past years. I wish we would not live all across the globe but nearby so that we could share delicious *yetsom migib* every now and then. I would like to make use of this opportunity to thank Rachelle for taking such good care of my children while I was writing the last chapters of this dissertation. Without you I would have not been able to concentrate so well on my work nor come home to such a relaxed atmosphere.

The last words of this acknowledgement are reserved for Solomon, my dear husband. We have walked an impossible road during the last years. You have always held my hand, no matter how high the mountain we had to climb or how deep the valley we had to cross, which shows what a wonderful strong and incredible sweet person you are. Our perseverance has been finally rewarded, it has brought us more than we could ever imagine, including two PhD dissertations, two amazing children and two people who have found true love at last. I am more than ready to live happily ever after with you and Helina and Nahom, wherever and however that may be. Thank you for all, thank you for being you.

Table of Contents

Table of Contents

1. An Introduction

1.1 Societal relevance: Simplicity on paper, complexity in practice?

Since the 1980s a major change took place in public policies for water resources management. The general approach in public policies shifted from an emphasis on physical water delivery by governments to creating an enabling environment for other parties to provide and use water resources. Whereas before governments primarily invested in the development, operation and maintenance of water infrastructure and were mainly concerned with the distribution of water, in the new approach they mainly focus on managing water resources systems by stipulating frameworks for water allocation (Cleaver and Elson, 1995; Allan, 1999; Neubert et al., 2002; Mosse, 2004; Lowndes, 2005; Swatuk, 2008; Saleth and Dinar, 2005; Mosse, 2006; Ahlers and Zwarteveen, 2009; Sehring, 2009). The mainstreaming of this substantial shift in the responsibilities of governments in the provision of services is largely the result of restructuring of the global economy in the aftermath of the global recession in the early 1980s (Sachs et al., 1995; Stiglitz, 2012)[1]. In the water sector this policy shift has been mainly consolidated and legitimized through deliberations between supranational organizations at a series of global water forums during the last three decades and is since actively disseminated through programmes of the World Bank and other funding agencies (Mosse, 2004; Ahlers, 2005; Conca, 2006; Molle, 2008). Governments who adopted these new public policies revised their water legislation and took up primarily an oversight role in the water sector. Through regulatory frameworks, organizational blueprints and specifying key principles they attempt to steer and control institutions that govern decision making over distribution, access and use of water resources at national, regional and local level. Rather than directly manipulating water resource configurations[2] through investments in infrastructural development, the bureaucrats became involved in crafting an institutional change process in the hope that it would lead to specific material outcomes aligned with their political ideals and ambitions envisioned in the policy reform process. But how does this shift in policy approach materialize in practice and how does it affect water resource configurations within river basins?

In their very essence policies are always based on simplified models of reality. This creates tension between the inevitable simplicity of policies on paper and inherently complex practice that they aim to steer (Long, 1989; Mosse, 2004; Lowndes, 2005; Lewis, 2009, Peck and Theodore, 2010; Bourblanc, 2012). As a result, and often to the disappointment of policy makers, policies seldom fully achieve the envisaged objectives and regularly have unintended consequences (Lowndes, 2005; Saleth and Dinar, 2005; Streeck and Thelen, 2005). However, according to a growing body of literature, a more fundamental issue is at stake within the conventional approach to policy that has been implemented since the 1980s. This mainstream approach is based on the assumption that institutions, here defined as the rules in use, can be

---

[1] It should be noted that this new approach was already experimented in the Chilean water sector since the 1970s, see Ahlers (2005) for a detailed analysis.
[2] In this dissertation I define water resource configurations as the materialized division in control over, access to and distribution of water between water users sharing the same water resource. With this definition I want to emphasize not only the social but also the historical and physical nature of the process through which water resource configurations are produced and maintained.

crafted through policy interventions, or in other words, it is assumed that institutions can be externally designed and optimized by policy makers and as such be implemented in practice (Ostrom, 1990; 1993; 1999; Saleth and Dinar, 2005). However, several scholars question this assumption and argue that the actual institutions that govern decision making in society are always hybrid in nature and thus seldom reflect solely the policy objectives (Cleaver, 2002; 2012; Mosse, 2004; Lowndes, 2005; O'Reilly, 2006; Peck and Theodore, 2010; De Koning, 2011). Policies are interpreted, renegotiated and rearranged at various spatial levels, a process closely intertwined with biophysical landscapes and uneven[3] social relations among actors. Institutions that result from this process will, to a greater or lesser extent, thus not only reflect the ambitions stipulated within policy frameworks but also configurations that are socially embedded at different spatial levels (Von Benda-Beckmann and Von Benda-Beckmann, 2006; Lowndes, 2005). Contrary to what policy makers might wish, actors do not solely strive for optimal resource use in this process, but also employ, and are circumscribed by, institutions that maintain or contest social consensus (Cleaver, 2002). Policy reforms will thus never be straightforward processes, especially when they specifically aim to alter institutions that govern society (Lowndes, 2005; Mollinga, 2008; Mosse, 2008; Swatuk, 2008). These contentious and ambiguous processes explain why policies so often lead to different outcomes than envisioned on paper.

Not only do scholars question the extent to which institutions can be crafted, they also argue that the mainstream policy approach has led to the proliferation of particular policy objectives for creating the enabling environment for (water) service delivery. They argue that the 'roll-back' of state services from provider to manager and the 'roll-out' of specific policy prescriptions to aim to craft 'optimal' institutions for the use of resources is brought forth by neoliberal political ideologies (Tickell and Peck, 2003; Harris, 2009). Without going into detail and acknowledging that neoliberalism cannot be seen as a single ideal or coherent policy (Jessop, 2002; Peck, 2004; Bakker, 2007), the basic consensus within the neoliberal paradigm is the supremacy of market rule in distributing resources efficiently and maximizing profit (Bakker, 2002; 2003; Harvey, 2005; Ahlers, 2005; Swyngedouw, 2009; 2011; Harris, 2009). Despite sometimes divergent ideologies and alternative ambitions of policy makers at national level, the strong involvement of supranational organizations (e.g. technocratic research organizations, UN agencies, World Bank) in policy making processes has led to the mainstreaming of policy prescriptions that creates an enabling environment in which the neoliberal project can unfold within different realms and at various localities (Burawoy, 2000; Ahlers and Zwarteveen, 2009; Budds and Saltana, 2013; Harris, 2009). Within the water realm, widely adopted policy prescriptions that are associated with neoliberalization of water include the decentralization of water management responsibilities to water users, the economization of water use through the introduction of cost-recovery fees, and the individualization, and in some cases privatization, of land and water rights (for full discussion see Bakker, 2000; Tickell and Peck, 2003; Ahlers, 2005; Harris, 2009). Concerned with the implications of neoliberalization, several scholars have pointed out how it has led to the exacerbation of structural inequities in societies across the globe[4] in terms of access to and control water resources (Ahlers, 2005; Bakker, 2005; Boelens and Zwarteveen, 2005; Harris; 2005; Hart, 2006; Bond, 2006; Bakker, 2007; Swatuk, 2008; Ahlers and Zwarteveen, 2009; Kemerink et al., 2013). Especially for African countries, with their strong dependence on supranational organizations and overseas development agencies for financial support as well

---

[3] In this dissertation I use the word "uneven" to refer to not only dissimilar but also inequitable, and as such contested, circumstances.

[4] See Stiglitz (2012) for a detailed analysis why markets failed to distribute resources efficiently, how markets reinforce structural inequities in societies and what the implications are for economies around the world.

as their limited human resources and high inequalities in distribution of wealth as result of the colonial history, this mainstream policy approach might have tremendous implications on water resource configurations, negatively affecting large sections of the population (van Koppen and Jha, 2005; Bond, 2006; Swatuk, 2008; Manzungu and Machiridza, 2009; Manzungu, 2012; Kemerink et al., 2013; Van Koppen and Schreiner, 2014; Kemerink et al., *forthcoming*).

In response to the criticism on the mainstream approach in public policy regarding natural resources, critical institutionalism has emerged as a school of thought which aims *"to understand how institutions work in practice and consequently why the outcomes benefit some people and exclude others"* (Cleaver, 2012:1; see also Cleaver and De Koning, 2015). Building on theories of critical social justice and political ecology and drawing from post-structural perspectives, critical institutionalism brings together scholars from different disciplines who encourage rethinking of key assumptions underlying the mainstream approach and offer alternative views on the institutions that mediate the relationships between the natural and social realms. Critical institutionalism has a fundamentally different conceptualization of what institutions are, how they emerge and endure, and how they shape human behaviour and (water) resource configuration than mainstream institutionalism. It allows for an institutional analysis approach that engages with the ambiguity, partiality and dynamics of institutions governing natural resources. However, critical institutionalism is criticized for its limited policy purchase as it fails to offer clear direction for bureaucrats (Blaikie, 2006; Mosse, 2006). According to the mainstream approach design principles can be selected by policy makers to optimize resource use, for instance in terms of efficiency, equity and/or sustainability. Hence, the impression is given that institutions governing resource configurations can be aligned with the political ambitions of the policy makers. However, critical institutionalism so far does little to set clear guidelines for policy makers on how to approach reform processes and as such does not reduce the uncertainty policy makers have to deal with (Cleaver, 2012; Cleaver and De Koning, 2015). Critical institutionalism currently mainly raises questions for policy makers without providing answers, such as: to what extent, how, and why can public policies steer institutions that shape water resource configurations? How to maintain responsiveness to local dynamics within public policies at national level? How to facilitate processes of progressive change to address structural inequities in access to water?

Being concerned with equity in water resource configurations, this study engages with critical institutionalism and examines the interaction between public policies adopted and implemented by government agencies and the institutions that govern access to, control over and distribution of water resources used for agriculture. How this interplay works out within waterscapes that are historically constituted by natural and social processes is the object of this dissertation. I do this by analyzing case studies in four African countries that have reformed their water policies during the last decades, namely Kenya, South Africa, Tanzania and Zimbabwe. With this research, I aim to provide the much needed insight for bureaucrats to understand the working and implications of current public policy approaches and seek to offer them more concrete directions for revisiting these processes within the water realm. Moreover, this research aims to contribute to advance the emerging theory on critical institutionalism by applying it to empirical cases and linking it with theories that illuminate constitutive spatial and material processes. Because institutional development is such a central focus of water policy since the 1980s, this chapter first provides conceptualization of institutions as well as policies and examines the interplay between them to better comprehend institutional change processes. This is followed by a deliberation on how material artefacts and natural processes shape institutional change processes and vice versa. Thereafter I will define the overall research objectives and research questions and explain the methodology and methods used in conducting this research. This chapter concludes with an outline of the remaining chapters of this dissertation.

1.2 Scientific relevance: complexity on paper, simplicity in practice?

1.2.1   Conceptualizing institutions

The mainstream[5] school of thought for understanding institutions is based on new-institutionalism, a theory that assumes amongst others that institutions can be crafted. Within this theory, as explained in the earlier works of Elinor Ostrom[6], institutional crafting is regarded as a continuous evolutionary process of developing the optimal institutions for interactions between individuals as well as between individuals and common pool resources (Ostrom, 1990; 1993; 1999). It argues that institutions can be externally designed and locally crafted following certain principles to achieve a shared goal, namely sustainable management of the resource. As such it is assumed that institutional formats are not only available, but also implementable and desirable for all actors. Without these institutional frameworks, actors are assumed to maximize resource use for their own benefit without considering other users or the conservation of the resource. Within this school of thought, institutions are thus conceptualized as human produced constraints and opportunities within which individuals can make choices and which shapes the consequences of their choices (McGinnis, 2011). In this way, institutions are assumed to provide individuals the security that others will act in agreed ways or otherwise be sanctioned, which stimulate them to cooperate for mutual benefit. Because of the emphasis on tangible and identifiable behaviours and incentives, there is a focus on bureaucratic institutions[7] based on explicit organizational structures and clear

---

[5] Albeit being aware of the partiality and ambiguity of this label, in this dissertation I use the term 'mainstream' to refer to the established, widely accepted and/or conventional understandings, approaches or practices in comparison to alternative perspectives, which (partly) have emerged in critique to these mainstream notions. These divergent understandings, approaches or practices I will refer to as 'critical' (see also paragraph 1.4.2 for a more detailed description of critical social theory).

[6] Even though I will critique throughout this thesis the simplistic view on institutions as put forward by new-institutionalism school of thought, I find it important to provide the context of Ostrom's work. Her research was inspired by her criticism on 'the tragedy of the commons' (Hardin, 1968) in which it is assumed that individual rent seeking behavior would deteriorate common pool resources (i.e. natural resources from which users cannot easily be excluded nor can be consumed by multiple users simultaneously such as water, pastures, forests) because of the disparity between the flows of benefits and costs for overexploiting these resource. Hence, it was suggested, to avoid a tragedy, these resources should either be privatized or controlled by the government. However, in her search to solve the collective choice dilemma, Ostrom showed with her research on labor intensive irrigation systems in Nepal that, given the right circumstances, communities are capable to collectively manage common pool resources. Based on her empirical research she identified eight 'design principles' to craft institutions that would facilitate sound management of common pool resources by collectives of resource users. Ostrom's work has become influential because it was picked up by policy networks to scientifically justify an already ongoing, political motivated, change in their policy approach (see also section 1.2.2). Her later work, in which she shares a more complex view on institutions albeit still emphasizing the need for explicit rules and direct incentive systems, has been largely ignored by the same epistemological community of policy makers.

[7] In this dissertation on purpose I chose to avoid the labels 'formal' and 'informal' for distinguishing between institutions that are sanctioned by the government and institutions that are not authorized by the government. In my opinion this kind of categorization would create a false dichotomy as institutions are often hybrid in nature and originate from various 'informal' and 'formal' sources and as such are often only partially sanctioned. Moreover, what is regarded as formal in a society depends on the legitimacy given to different kinds of

delineation of resource use (Cleaver, 2002; 2012). Underlying this theory is the concept of rational choice in which it is assumed that individuals make rational decisions based on *"the benefits and costs of actions and their perceived linkage to outcomes that also involve a mixture of benefits and costs"* (Ostrom, 1990: 33). In this view it is assumed that human agency, which can be broadly understood as the capability of actors to choose and to act, is only bounded by incomplete information necessary to take strategic actions. Feminist scholars have deconstructed this 'separate self model' in which individuals can act unhindered by their social, material and political context. They have shown how this narrow model of human beings ignores historic inequities and contemporary social struggles that shape human agency. They argue that a level playing field does not exist: actors cannot interact freely as they are always bounded in their actions by uneven social relations or unequal access to resources (Folbre, 1994; 2012; Elson, 1995; 2012; Beneria, 1999; 2004; Ahlers, 2005; Zwarteveen, 2006; 2011; Ahlers and Zwarterveen, 2009). Privileging single aspects of people's identities for policy purposes is thus problematic as the concept of rational choice falls short in recognizing humans as social beings with multiple social identities and complex webs of affiliations that shape their behaviours and circumvent their actions (Cleaver, 2002; 2012; O'Reilly, 2006). As result of this shortcoming, I argue that new-institutionalism poorly conceptualizes institutions and therefore fails to explain the ambiguity, partiality and plurality of institutions and thus the context specificity of institutional change (see also Giddens, 1984; Long and van der Ploeg, 1989; Cleaver, 1999, 2002; 2012; Boelens, 2008; Molle, 2008; Ahlers, 2010; Laube, 2010; De Koning, 2011; Kemerink et al., 2013; Komakech et al., 2012b).

A more nuanced view is articulated by theories that I here broadly refer to as critical institutionalism (Cleaver, 2012; Cleaver and De Koning, 2015). Coming from different disciplines and having various foci and nuances, a common understanding within critical institutionalism is that both agency and social structures shape human action. This dual view on human behaviour builds further on the earlier works of Giddens in which he argued that actors always have some degree of agency, even under the most oppressive conditions and even if only through mundane sanctioning processes of *"disapproval, criticism or simply an absence of response"* (1984:175), yet also are always bound by some level of subordination. This expresses the reciprocal albeit unequal relations of autonomy and dependence between actors. Institutions emerge from as well as shape these relations (see also Long, 1984; Long and van der Ploeg, 1989; Long, 2001). Within critical institutionalism institutions are thus conceptualised as outcomes of dynamic social processes in which authority is constantly contested, negotiated and reaffirmed, and can be defined as the rules is use that shape, regulate and reproduce human behaviour across time and space (Mollinga, 2001; Cleaver, 2002; 2012; Boelens et al., 2005; Von Benda-Beckmann and Von Benda-Beckmann, 2006; Boelens, 2008; Molle, 2008; Ahlers, 2010; Laube, 2010). Critical institutionalist scholars are concerned with understanding the social processes through which institutions emerge and endure. For instance, Cleaver (2002) explains how institutions for collective management of (water) resources *"are formed through processes of bricolage in which similar arrangements are adapted for multiple purposes, are embedded in networks of social relations, norms and practices and in which maintaining social consensus and solidarity may be equally important as optimum resource management outcomes."* (Cleaver, 2002:17; see also Douglas, 1987). Processes that she calls institutional bricolage thus elude the design principles commonly propagated within the new-institutionalism theory. Instead, processes of institutional

---

authorities, including but not limited to the state government, which might change over time and vary across space and which might be perceived differently by disparate actors (see also Cleaver, 2002).

bricolage show how institutions emerge through daily interactions and improvisations building on existing institutions and styles of thinking and therefore are deeply embedded in sanctioned social relationships and everyday practices. As a result institutions may work intermittently and in an ad hoc manner, though nevertheless be enduring and approximately effective (Cleaver and Toner, 2006; Cleaver and Franks, 2007; Cleaver, 2012; Komakech et al., 2012a). Within this process actors, referred to as 'bricoleurs', are seen as both rational and social human beings who are *"deeply embedded in their cultural milieu but nonetheless capable of analysing and acting upon the circumstances that confront them"* (Cleaver, 2002:16). They, consciously and unconsciously, rework institutions borrowing from past and present rules and practices forming hybrid patchworks of institutional arrangements.

Scholars who study the anthropology of law come to similar conclusions on the hybrid nature of institutions originating from various temporal and spatial sources. In their aim to understand the social processes through which constellations of institutions emerge, maintain and change, these scholars analyze the coexistence and interaction of different normative orders in the same social-political space that govern human interaction (Von Benda-Beckmann, 1997; Von Benda-Beckman and Meijl, 1999; Boelens et al., 2005; Von Benda-Beckmann and Von Benda-Beckmann, 2006; Kemerink et al., 2011). In this analytical approach, generally referred to as legal pluralism, normative orders can be understood as any system of rules or shared expectations of what people should or should not think, say or do concerning a particular situation imbued by world views. This moves law beyond state-recognized legality and encompass other possible forms of institutions derived from normative orders that may originate from various sources such as political ideologies, economic dogmas, knowledge regimes, religions and cultures at different spatial and temporal scales. Norms are thus articulated and materialize through institutions that shape human behaviour and interaction (Boelens, 2008). The different normative orders in society can be complementary, overlapping or even contradictory creating space for bargaining and manipulation by different actors: *"actors all draw on legal repertoires, interpreting and using them in the pursuit of their interests"* (Von Benda-Beckmann and Von Benda-Beckmann, 2006:10). Nevertheless, within legal pluralism the dual conceptualization of human behaviour (Giddens, 1984) is also acknowledged as actors are not only assumed to consciously 'shop around' for normative orders through which they can best exert their agency, but also are constrained by socially 'accepted' normative orders imposed by others (Von Benda-Beckmann and Von Benda-Beckmann, 1997; 1999; 2006; Boelens et al., 2005; Meinzen-Dick and Nkoya, 2007). How legal constellations play out in social life and generate a plethora of hybrid local rules and arrangements is thus in its very essence shaped by history and embedded in local realities.

Like critical institutionalist scholars, I am particularly concerned with unravelling how social processes of institutional change produce different outcomes for diverse social groupings. It is commonly understood that actors with stronger leverage positions as result of uneven access to material resources have a greater influence on what does or does not happen in society. They can maintain, even though never absolute, their authority through various means of control despite resistance and struggle. But what is the role of institutions in this process, how are structural inequities maintained and contested over time? Whereas critical institutionalist scholars employ different notions of power, I find it useful for this research to adopt a Foucauldian notion in which power is not necessarily only possessed and exercised by actors, but also operates in the invisible space of what we leave unquestioned, that what we have internalized and taken for granted (Foucault, 1979; 1980; Haugaard, 2002; Mills, 2003; Ekers and Loftus, 2008). This notion allows us to deconstruct how power works through the existence and proliferations of norms that stipulate what is regarded 'right' and 'wrong', and for whom, beyond specific contexts and beyond certain eras (see also Scott, 1986; Boelens, 2008; Zwarteveen, 2008). The interests actors pursue, and the normative frames they draw on, are closely intertwined, subsequently reproducing hegemonic normative frames, while alternative normative understandings in society are dissuaded. This is neither a straightforward nor a neutral process, but highly political as uneven relations of power become embedded in broader forms of dominant social, cultural and economic structures (Foucault, 2000a; Ekers and Loftus, 2008). Building on feminist political ecology, Nightingale (2011) for instance argues that "... *regardless of their historical origin, the repetition of normative social identities is crucial to the production of subjectivities as it is through these discourses and the internalisation and contestation of them that the subject is (violently) achieved*" (Nightingale, 2011:155; see also Foucault, 1980; Butler, 1990). In his conceptualization of power Foucault emphasizes two meanings of the word subject, namely "*subject to someone else by control and dependence, and tied to [an actor's] own identity by a conscience or self-knowledge*". He continues with stating that "*both meanings suggest a form of power which subjugates and makes subject to*" (Foucault, 2000b: 331). He thus argues that not only stronger actors enforce and reproduce subjectivities, but also the subjects themselves become involved in processes of what he refers to as 'subjectification' through internalizing truth claims and normative understandings of reality and conforming to uneven institutions that govern society. Similarly, Nightingale argues that, through the continuous (re)production of social difference in everyday practices, "*subjectivity can be a contradictory achievement with subjects exercising and internalizing multiple dimensions of power within the same act*" (Nightingale, 2011:155; see also O'Reilly, 2006). Power is thus conceptualized as various forms of relational means that function, at least partly, through the presence and proliferation of norms within networks of relationships upheld by both the dominant and subordinated actors (Boelens, 2008). Through this complex normalizing process, subjective social relations become over time embedded in unconscious routines and ritualized ways of doing, including the ways in which actors perceive themselves, others, and the social and material reality around them. Bourdieu (1977) therefore argues that to understand social relations we need to unravel everyday practice of actors within the context of time and space (see also Van der Zaag, 1992). He explains that the context of time is relevant as the actions actors take are constituted by former practices and experiences of the actors and as such practices are inherently historical (see also Cleaver, 2002) and the context of space is relevant as the actions of actors always take place within a physical and material environment that shapes their practices.

In daily practice uneven institutions materialize, producing social differences among actors. Scott (1986) argues that social differences are produced by giving meaning to perceived biological differences and/or through internalization and embodiment of norms, for instance norms on how somebody should dress, talk or walk. I regard social difference problematic when mobilized to signify, reproduce and consolidate subjective relationships and/or when used to legitimize structural material inequities in society (see also Scott, 1986; Nightingale, 2011; O'Reilly, 2006). Gender, race, ethnicity, age and class have become persistent constitutive elements of subjective relations that affect all actors in society to "*the extent that these ... establish distributions of power (differential control over or access to material and symbolic resources) ... [and thus] becomes implicated in the conception and construction of power itself*" (Scott, 1986:1069). These constitutive elements of social differences intersect, creating for instance not only dichotomies between men and women, but also between black and white men and between young and elderly women. It is this intersectionality of the constitutive elements of social difference that shapes the social identities of actors (Burman,

2004; O'Reilly, 2006; Valentine, 2007; Ahlers and Zwarteveen, 2009; Nightingale, 2011). These social identities prescribe actors particular normative behaviours, bounding their actions and shaping their interactions with other actors. As the production of social difference is a continuous and contested process, the idea of fixed or universal identities can be questioned (Nagar, 2000; Gibson, 2001; O'Reilly, 2006). For the case of gender, Scott (1986) therefore argues that social categories such as 'man' and 'woman' are at once empty and overflowing: *"Empty because they have no ultimate, transcendent meaning. Overflowing because even when they appear to be fixed, they still contain within them alternative, denied, or suppressed definitions"* (Scott, 1986:1074). Also portraying 'woman' and 'man', 'black' and 'white', 'rich' and 'poor' as inherently binary or even opposing categories is problematic as it does not recognize interdependencies and complementarities between them that also exists along conflicts and struggle[8] (Scott, 1986; Cleaver, 1999; Ahlers, 2009). The social identities of actors are thus complex, ambiguous and might change during their life courses, yet they are at the same time also deeply embedded in the prevailing normative frames of the society they live in (Cleaver, 1999; 2012). I take from this conceptualization of social identities that actors' agency and social constraints are neither rigid nor equal, but dynamically shape the actors' choices and ability to act, including their capability to respond to policy interventions and/or to manipulate institutional change processes.

Even though critical institutionalism has proven useful in understanding how institutions emerge, endure and change, especially at local level, and how institutions produce differential outcomes for actors (Cleaver, 2012), it also leaves questions unanswered. For instance, how do institutions mediate between the social and the material? How does materiality of natural resources shape institutions? How do ecological processes affect institutional change? What is the role of material artefacts such as infrastructures in these processes? In other words, how do the agencies of non-human nature constitute and change institutions? I shall return to these questions in section 1.2.4 of this dissertation, but first I will discuss another issue that in my view is not yet sufficiently incorporated within critical institutionalism: the implications of the global-local continuum in terms of constitutive processes that dynamically connect various geographical scales and produce similar water resource configurations in different geographical locations (Conca, 2006; Hart, 2006; Harris, 2009; O'Reilly et al., 2009; Swyngedouw, 2011). Hart argues that struggles over resources are local articulations of forces at play in national and international arenas and therefore *"divergent but increasingly interconnected trajectories of ... change ... are actively constitutive of processes of globalization"* (Hart, 2006: 981). It is therefore crucial to understand the interplay between global structural forces and local historic particularities. Where critical institutionalism is well suited for analyzing contextualized local institutional arrangements, it currently pays less attention to structural configurations of institutional processes at larger spatial levels and how these configurations interact with the institutional arrangements at local level. The structural forces can be directly linked to the current capitalist state of the global political economy[9]. Not only does the capitalist mode of production lead to particular material outcomes, it also reveals an ongoing isomorphic process through which similar institutions manifest themselves

---

[8] Throughout this research I analyze the data using disaggregated social categories based on gender, race, ethnicity and class. I realize that this might contribute to the reproduction of simplistic and stigmatizing social identities. However, I do so particularly to show the diversity of actors within such categories and to discuss the interrelations and dependencies between various kinds of actors within society.

[9] In this dissertation I will refer to the capitalist political economy and more specifically the neoliberal ideology. However, I will not in great detail discuss the ontology of the economic system nor use the political economy approach to analyze the capital and resource flows within the case study countries. Rather I take the current political economy in Southern Africa as the context in which the water reforms have been produced, enacted and implemented (Swatuk, 2008) and reflect on how this shapes the outcomes of the reform processes.

at different places around the world. This, after all, might indicate that institutions do not fully elude design and potentially points to the concentration of agency in the hands of a few influential actors operating at a supranational scale who are actively, and effectively, involved in shaping institutional change processes. To understand the implications of this global-local continuum and how it works out within the water realm, I shall now turn to theories that try to explain the perceived homogeneity in institutions by conceptualizing contemporary policy-making processes and how it leads to the persistence of particular policy models.

1.2.2 Conceptualizing policies

Policies are developed to structure and justify decision making and guide interventions. Policies can be understood as overall plans or strategies that stipulate established principles, general goals and prescribed procedures of the organization on the issue at stake (Lodge and Wegrich, 2005). These policies may be explicit but can also be more implicit or largely symbolic in nature (Kemerink et al., 2012). Even though policies are sometimes regarded as institutions, I find it useful to distinguish between the two concepts as institutions emerge from interactions as well as govern these interactions, while policies are consciously designed to, directly or indirectly, manipulate institutions but not necessarily do so.

Similar to the different views on institutions, also the conceptualization of public policy differs between schools of thought that base their assumptions exclusively on the concept of rational choice and those who reject this narrow conceptualization of human agency (Mosse, 2004; Griggs, 2007). Within the first school of thought policy-making is regarded as a linear, or at most iterative, process that runs through neatly defined successive phases of problem identification, policy formulation, policy implementation and evaluation of its impacts (Brewer and DeLeon, 1983; DeLeon, 1999; Griggs, 2007; Jann and Wegrich, 2007). The policy makers, who ought to *"contribute to problem solving or reduction of the problem load"* (Jann and Wegrich, 2007:53), are assumed to be neutral and capable to disconnect from their own interests and perceptions within the policy-making process (Andrews, 2007), only bounded in their actions by insufficient data. Underlying this school of thought is a positivist epistemological framework that postulates that certain problems can be solved if 'objective' and 'valid' scientific knowledge is provided (Kornov and Thissen, 2000; Rap, 2006; Conca, 2006; see also paragraph 1.4.1) assuming that *"scientific knowledge is a key input that contributes to the best outcome"* (Andrews, 2007:162).

Rejecting this simplistic view, I concur with scholars who emphasize the political nature of the policy-making process in which policy making is seen as *"a social process of and between actors, rather than a rational effort to search for the optimal solution given a fixed problem definition"* (Hermans and Thissen, 2009:808). These scholars argue that policies are outcomes of a discursive practice of policy making in which problems are framed and ideas, concepts and categories are aggregated through which meaning is given to a particular phenomenon (Hajer, 1995; Mollinga 2001; Griggs, 2007; Peck and Theodore, 2010). Within this school of thought policy makers are not considered objective and solely rational and the subjectivity of scientific knowledge is emphasized. As Rap argues, policy making is *"... an ongoing process that transcends the artificial boundaries between politics, bureaucracy, and research and the neat stages of policy formulation, implementation, and evaluation. Researchers, consultants, and advocates can play a significant role in the advancement of a certain interpretation of policy"* (Rap, 2006:1304). Within the process of the proliferation of a policy, epistemic communities or expert networks gradually establish, sharing ideological understandings and cultural practices (Conca, 2006; Rap, 2006; Molle, 2008; Peck and Theodore, 2010). Rhodes defines these networks as *"sets of formal institutional and informal linkages between governmental and other actors structured around shared if endlessly negotiated beliefs and interests in public policymaking and implementation. These actors are interdependent and policy emerges from the interactions between them"* (Rhodes, 2006: 424). Similarly, Peck and Theodore (2010) state that *"policy actors are not conceptualized as lone learners ... and rarely do they act alone."* (2010: 170).

Several scholars argue that specific storylines, referred to as policy narratives, are influential within the policy-making process (Roe, 1991; 1994; Hajer, 1995; Mosse, 2004; Rap, 2006; Molle, 2008; Peck and Theodore, 2010). These policy narratives can be understood as specific and stabilized interpretations of physical and/or social phenomena that assume certain fixed causal relationships not necessarily grounded in empirical evidence: *"Narratives ... are often self-validating because they tend to produce evidence rather than the other way around"* (Molle, 2008:137). As an example he mentions the nowadays popular policy narrative that waste of water resources is the result of the lack of pricing to reflect the real costs of using water. This narrative legitimizes charging cost-recovery fees for water, which might lead to a reduced use of water and thus produces 'evidence' of a positive correlation between the cause and the effect. However, in reality this simplistic narrative obscures the complexity of the processes at play and the unequal options actors have: for instance in the case study in Kenya, I will show how small-scale farmers who pay for water but depend on a collective inflexible water distribution system are not able to optimize their water use, while other small-scale farmers in the same catchment, who can access water on an individual basis by pumping straight from the river, use water more efficiently despite the fact that they do not pay any fee for the water they use (see chapter 5). Nevertheless, even when confronted with contradicting empirical studies, the narratives maintain and tend to *"continue to underwrite and stabilize the assumptions"* (Roe, 1994:2) for policy-making. Molle (2008) relates the persistence of policy narratives to ideological underpinnings of policy networks, who articulate their normative views through what he calls nirvana concepts. He defines nirvana concepts as the embodiment of *"an ideal image of what the world should tend to ... They represent a vision of a 'horizon' that ... societies should strive to reach"* (Molle, 2008: 132). This metaphor emphasizes not only the inherently ideological origin of policies, but also the intrinsic future-oriented perspective of policies. Molle continues that, even though the chances that nirvana may be reached are admittedly low, *"the mere possibility of achieving them and the sense of 'progress' attached to any shift in their direction suffice to make them an attractive and useful focal point"* (Molle, 2008: 132). The persistence of policy narratives can be seen as the result of the continuous support of a policy network to validate these narratives because their ideal image of the future is constructed on the assumed causality embedded in the narrative (Latour, 1996; Rap, 2006; Mosse, 2004; Molle, 2008; Peck and Theodore, 2010).

The policy narratives are believed to produce and legitimize certain 'paths towards nirvana' in the form of policy models. Rap defines these policy models as *"particular, stabilized interpretations of policy-related events that is used to generate similar policy in other parts of the world"* (Rap, 2006:1302). As such policy models *"seek to stabilize and validate an explicit set of rules, techniques, and behaviours, that when applied in 'foreign' settings might be expected to yield comparable results"* (Peck and Theodore, 2010: 170; see also Rusca and Schwartz, 2012). Policy models are widely embraced by governments and development agencies who prefer working with simplified models as they are *"apparently sanctioned by experience, approved by experts and powerful institutions, and using them seemingly minimizes risk"* (Molle, 2008:138; see also Roe 1991; Uphoff et al., 1998; Cleaver, 2002;

Mosse, 2004; Rap, 2006; Laube, 2010; Peck and Theodore, 2010; Bourblanc, 2012). Moreover, policy models fit well with the positivist aims for 'objectivity' and 'neutrality' that are dominant within the development orthodoxy as it assumes that performance of the standardised policy can be simply measured and compared between sites, and as such interventions can be justified in parliament, based on predefined indicators (Power, 2000; Rap, 2006; Peck and Theodore, 2010). However, not only does the use of policy models ease the procedures of government agencies, there also seem more strategic reasons why generic policy models are so popular within the contemporary policy making processes. Based on Haas (1992) Rap describes how policy models are *"subject to a continuing process of production and promotion aiming to mobilise and maintain political consent among the epistemic community to which they are directed and which they shape"* (Rap, 2006:1304). With careers and other personal gains closely tied up with the adoption of a particular policy model, the policy makers involved in the production and promotion of the policy model are believed to do so to maintain authority and to pursue, or at least protect, their own interests (Allan, 1999; Mosse, 2004; Lowndes, 2006; Rap, 2006; Molle, 2008; Peck and Theodore, 2010; Budds and Sultana, 2013). Furthermore, Conca argues that expert networks with particular value orientations, through circulation of narratives and pressuring governments to adopt policy models, have become an *"authoritative source of norms in world politics"* (Conca, 2006:126; see also Goldman, 2007). Similarly, Peck and Theodore (2010) claim that *"the quasi-academic trappings of the World Bank Institute, its intellectually colonizing 'knowledge bank' strategies, and the widespread concern with 'scaling-up' favored projects can all be seen as a manifestation of a certain kind of normative authority"* (2010:171). This normative authority may be explicitly exercised through for instance conditionalities attached to funding sources or through regulatory frameworks, but it may also be more implicitly enforced through informal pressure within hierarchal organizational structures (DiMaggio and Powell, 1983; Lodge and Wegrich, 2005; Frumkin and Galaskiewicz, 2004).

It is within this political understanding of policy-making processes that I will analyze the public policies that are pursued within the water sector reforms in eastern and southern Africa, with a particular focus on agricultural water use. I acknowledge that policy making is a highly dynamic process and at any point in time several (overlapping) policy networks may exist at different spatial levels. These policy networks might have different normative views and aim to pursue different interests within the same policy domain and as such compete for authority. After all *"... hegemony ... is an always incomplete process. The powers of network-normativity and model-making maybe be formidable, but they are far from totalizing, since they are also marked by contradiction and contestation"* (Peck and Theodore, 2010:171, see also Foucault, 2003). This contested process leads to continuous changes in the content of policies as well as to differences in policies at various locations. Nevertheless, within the reform processes ongoing in the countries selected for this dissertation, I observe striking similarities in narratives used to justify the reform processes as well as the implementation of similar policy models in dissimilar contexts. Without going into detail in the genealogy and content of the integrated water resources management (IWRM) paradigm, it can be considered as a 'nirvana' that is strived for by a global policy network based on specific narratives and legitimizing associated policy models (Molle, 2008; see also Savenije and Van der Zaag, 2002; 2008; Swatuk, 2008; Van der Zaag, 2005; Conca, 2006; Anderson et al., 2008; Mtisi, 2011). The policy network around IWRM gradually established and started to share ideological understandings and practices with actors engaged in the neoliberal project such as international development banks, overseas development agencies, and technocratic research organizations. This confluence with neoliberal ideals has led to two major changes in the policy making processes within the water domain. First, it pushed the policy making process from a mainly national to a primarily global arena, strengthening the involvement of supranational actors and stimulating isomorphic behaviour (Conca, 2006; Goldman, 2007). And second, it shifted the content of water policies from a physical orientation to a focus on management and institutional processes[10] (Cleaver and Elson, 1995; Allan, 1999; Neubert et al., 2002; Mosse, 2004; Swatuk, 2005; Conca, 2006; Ahlers and Zwarteveen, 2009). As such, the IWRM inspired reforms carried out across the globe, at least partly, opened up the road to disseminate neoliberal narratives and roll out concurrent policy models that align with neoliberal interests such as decentralization, privatization, formalization and economization of water resources management (Bond, 2004; Smith, 2004; Harvey, 2005; Bakker, 2007; Goldman, 2007; Laurie, 2007; Ahlers and Zwarteveen, 2009; Harris, 2009; Manzungu and Machiridza, 2009). Beyond critiquing the neoliberal ideology, I am particularly interested in illuminating what happens *'between the monotheistic privilege of dominant policy models and the polytheism of scattered practices surviving below'* (Mosse, 2004:645 quoting De Certeau, 1984). How are the policy models seized, interpreted and rearranged by actors within their discourses, strategies and negotiations over water at local level? How do policies influence institutions and under which circumstances do they become constitutive elements of water resource configurations? And how do institutions at different spatial levels shape policies making processes as well as implementation practices and abilities? To answer these questions, we need to better understand the interplay between public policies and institutions, which will be further explored in the next section.

1.2.3   Conceptualizing the interplay between policies and institutions

Various scholars have tried to conceptualize what happens with public policies once they 'get implemented' in practice. For this dissertation I am particularly interested in understanding how public policies interact with existing institutions, and whether or not, to what extent, under which conditions and in which direction public policies can steer institutional change. Based on a vast literature review of policy science and building on adaptive governance theories, Huitema and Meijerink (2009; 2010) discuss the role of policy entrepreneurs[11] and windows of opportunity to analyze how policies can lead to institutional change (see also Kingdon, 1984; Lowndes, 2005). Particular moments in time, such as elections, crises or disasters, are assumed to offer opportunities for policy entrepreneurs to initiate and accelerate institutional change processes. Key challenge for policy entrepreneurs is then to recognize, open, expand and finally use these windows of opportunity. According to the authors policy entrepreneurs can be found anywhere as long as they are *"good advocates of new policy ideas and good policy brokers"* (Huitema and Meijerink, 2010:5). While I disagree with the depoliticized assumption of unconstrained entrepreneurship, I sympathize with their view that

---

[10] This shift has been accompanied by a worldwide decrease in the investments in the agricultural sector since the late 1970s, particularly effecting large-scale public irrigation schemes. This trend is especially noticeable in the money lent by the World Bank to national governments for irrigation development; end 1970s the total World Bank lending accumulated to more than 2 billion US dollars while early 2000s it had dropped to less than 0.2 billion US dollars (Faurès et al., 2007). Similarly, in their report NEPAD discusses how the agricultural sector in Africa has been weakened by two decades of simultaneous private sector and state disinvestment and reduced aid assistance for supporting irrigated agriculture. They estimate that post-colonial public investment in agriculture by the African countries dropped from its peak mid 1980s from 8% of the total government expenditures to nearly 2% by the end of the 1990s (NEPAD, 2010).

[11] Policy entrepreneurs can be sees as individuals who introduce, translate and push the adoption of new ideas into the public policy practice. They often operate outside the formal position of the government and remain largely on the background, however, they are believed to be instrumental in influencing political agendas, framing policy issues and steering policy debates (Roberts and King, 1991; Mintrom, 2000).

"change can perhaps not be managed in the sense of being preplanned and centrally controlled, but it can at least be prepared for and 'navigated' from point to point" (Huitema and Meijerink, 2010:4). The question then is who can prepare and navigate this process and in which directions? I find it useful to draw on the work by Lowndes (2005) who gives a more critical perspective on the agency and motives of policy entrepreneurs. She recognizes that unequal social relations at different spatial levels drive and constrain institutional entrepreneurship and she stresses that ideas and narratives promoted by influential policy networks manipulate the direction of entrepreneurship (see also Roe, 2004; Rap, 2006; Molle, 2008). Moreover, while Huitema and Meijerink (2009; 2010) assume the altruistic intentions of policy entrepreneurs, Lowndes emphasizes the partly self-centred motives when she states that *"institutional change depends critically upon the creative work of institutional entrepreneurs, who seek to expand and recombine their institutional resources as they face new challenges (and pursue, or at least protect, their own interests)"* (Lowndes, 2005:299). Even though she describes the crucial role of entrepreneurs to initiate institutional change and identifies different strategies that entrepreneurs employ in this process, Lowndes (2005) does not discuss in detail what happens in everyday life where public policies and socially embedded institutions intersect.

Within the critical institutionalism school of thought, De Koning (2011) describes three different kinds of bricolage processes that could happen when public policies are introduced into a particular setting in which existing institutions govern access to resources. The first process she calls 'aggregation' in which different institutional elements are combined. In this case, policies are to a certain extent adopted and combined with the existing institutions but not necessarily changing the essence of these institutions. The second process she identifies is 'alteration' in which the policy *"leaves a mark on the local institutional framework but does not achieve its original objective"* (De Koning, 2011:215). Within this process policies are altered, manipulated and partly incorporated, potentially changing existing institutions considerably. The last process she calls 'articulation' in which policies *"bounces off the shield of socially embedded institutions"* (De Koning, 2011:215)[12]. The effect of the policy is minimal, hardly even perceptible, and the existing institutions appear non-negotiable and unchangeable. This process might however lead to more plurality in society in which the normative frame of the policy coexists, and potentially clashes, with the normative frames of the existing institutions, creating room for manoeuvre and possibilities for forum shopping (Von Benda-Beckmann, 1999; Boelens, 2005; Von Benda-Beckmann and Von Benda-Beckmann 2006; Meinzen-Dick and Nkoya, 2007; De Koning, 2011). Even though the different bricolage processes De Koning identifies are useful to understand what could take place in the interaction between policies and institutions, in my view it does not yet illuminate why and under which conditions these different processes occur, nor does it reveal what the implications are for resource configurations.

Sehring (2009) uses the metaphor of corridors to analyze to what extent policy reforms lead to institutional change. Building on institutional bricolage, she situates what happens between institutions and policies as *"between path dependency and the development of new, alternative paths, which are never completely new but a recombination of existing institutional elements and new concepts"* (Sehring, 2009:66). In her work Sehring (2009) defines the institutional

---

[12] Personally I find the use of the word articulation in this context confusing as it could be interpreted as the action of being jointed together or the manner of interrelating. Hart defines articulation based on Hall (1985) *"the joining together of diverse elements in the constitution of societies structured in dominance"* (Hart, 2006: 998). Perhaps the word refutation or perpetuation might therefore better fit the processes De Koning (2011) refers to.

corridor as the room for manoeuvre available to actors to choose from a range of institutional settings. As such, the institutional corridor is delineated by the configurations of, and plurality in, the existing institutions as well as the extent to which the various actors are involved in decision-making processes. In this metaphor, narrow corridors are associated with institutional settings that offer little room for manoeuvre to a limited number of actors that are involved in decision making. Sehring (2009) argues that these narrow corridors will more likely lead to path-dependency in which historical experiences, policy legacies and existing institutional arrangements determine the outcomes of the reform process rather than the content of the new policy itself (see also Hall and Taylor, 1996; Thelen, 1999; Pierson, 2000; Lowndes, 2005). Thelen (1999) argues that the adherence to particular institutional paths is the result of feedback mechanisms with functional as well as distributional effects. Institutional path-dependency has a functional effect because *"once a set of institutions is in place, actors adapt their strategies in ways that reflect but also reinforce the 'logic' of the system"* (Thelen, 1999:392). Distributional effect of path-dependency refers to the uneven social relations in society that are reinforced by institutions that continuously marginalise actors who prefer alternative institutional arrangements. The institutional stickiness to path-dependency is thus not only assumed to perform certain functions, but it also serves certain interests (Pierson, 2000; Lowndes, 2005). Considerable societal investments in large hydraulic infrastructure also have been believed to aggravate path-dependency (Huitema and Meijerink, 2010). Through so-called 'lock-ins', in which everything supposedly must be geared towards maintaining the infrastructure to capitalize on the investment, path-dependency is legitimized (see also Molle, 2008). However, wider corridors, in which plural normative orders prevail and diverse groups of actors participate, alternative paths might emerge as wider corridors leave room *"... for more bricolage options ... [in which] the newly introduced ... rules have been adapted to existing institutional arrangements ... [and] traditional ... rules have been transformed to adjust to new conditions"* (Sehring, 2009:77). In other words, in wider corridors processes of alteration as described by De Koning (2011) can take place that potentially change the existing institutions considerably. Combining the two theories of Sehring (2009) and De Koning (2011) further explains how narrow corridors can become wider as a result of policy reforms: in narrow corridors processes of articulation might add 'new' normative frames underlying the policy reform to the institutional palette, thus widening the corridor and opening up opportunities for future change. Even though this combined conceptualization of bricolage processes in corridors gives insight why or why not institutional changes might happen in response to policy reforms, it does not reveal in which directions these changes happen and how it shapes particular resource configurations.

Drawing on legal pluralism and the theories on persistent policy models discussed in previous sections, I find it useful to also look at the normative frames underlying existing institutions as well as the proposed public policy reforms. This can potentially help policy makers and scientists to analyze why or why not and in which direction institutional change happens in response to policy reforms. We can hypothesize that if the metaphoric corridor is narrow it is more likely that the normative order underlying the policy reform is aligned with the normative order already prevailing in society as the few dominant actors most likely are tied up in policy networks that produce certain policy reforms. In this case, policies could then be easily adopted through processes of aggregation, reinforcing the dominant normative frame. If the corridor is 'narrow' but the normative order underlying the policy reform is considerably different from the normative order prevailing in society, for instance through external pressure or conditionalities of funding agencies, the policy most likely would not lead to considerable institutional change as a result of processes of articulation. In this case the institutional changes will most likely be limited to merely ceremonial changes (DiMaggio and Powell,

1983) though perhaps it increases the legal plurality in society providing room for other actors to become involved in decision making and as such widening the corridor over time. If the metaphoric corridor is wide enough the policy reform could be tweaked, reworked and adopted by various actors through processes of alteration. In this case the outcomes remain uncertain, though we can understand that also here analyzing the underlying normative frames can be informative as the policy reforms most likely reinforce those institutions with which it shares its normative underpinning. Of course this remains a very abstract conceptualization of the interplay between policies and institutions and questions arise such as how to 'measure' the width of an institutional corridor and how the constitutive processes of institutional change and emergence of an institutional corridor are initiated, what is the chicken and what is the egg? Nevertheless, what this abstract conceptualization does give us is that unravelling the normative orders in society as well as those underlying policy reforms helps to understand the interplay between institutions and policies and provide insight in how they shape water resource configurations. It is important to also note that policy reforms may originate from multiple normative orders which further complicate the interface with existing institutions, resulting sometimes in ambiguous and contradicting outcomes. For instance in the case study in South Africa I will argue that the internationally praised water act is based on conflictive neoliberal as well as socialist oriented ideology and show how this hampers the implementation of particular parts of the reform process (see chapters 3 and 4). Moreover, policy reforms might simultaneously take place at multiple levels and in various domains that potentially intersect, leading to cross-fertilization or clashes between normative frames (Lowndes, 2005). For instance in the case study in Zimbabwe I will discuss how the sudden change in the normative notions guiding the land reform process affected the implementation of the water reform process (see chapters 6) .

Based on the above, the question we are then left with is: how to detect norms? As stated before they are fundamental believes in society of what people should or should not think, say or do concerning a particular situation rooted in specific world views. Normative perspectives might be expressed through clear rhetoric, though they might also be left unstated or even be actively concealed. Nevertheless, whether implicitly or explicitly stated, prevailing normative orders materialize in practice through the interactions between people and between people and the environment while they carry out their daily activities, whether that be fetching water, irrigating land, authorizing water permits or collecting water fees (see also Bourdieu, 1977; van der Zaag, 1992). Within waterscapes normative frames underlying policies and institutions do not only intersect with each other, but also with physical processes and historical distribution of natural resources. The ultimate outcomes of this complex process will eventually become inscribed on the landscape and as such landscapes, or in this research waterscapes, *"...embody layer upon layer the legacies of former institutional arrangements, and of the changing environmental entitlements of socially differentiated actors"* (Leach et al., 1999:239). Hence, to unravel how normative orders shape water resource configurations and vice versa, I thus need to also study the material artefacts and physical processes within the waterscape.

1.2.4   Understanding waterscapes

"Irrigated landscapes and the ecologies they produce long outlive the particular alliances that created them ... Because water moves, it erases as well as makes social boundaries; it changes landscapes, provides the basis for new claims and threatens established orders ... Yet, it is also true that, as in Africa, where water development works were used to assert land

claims, influential or richer farmers can ensure that it is their rights that are fixed in permanent concrete structures, such that the technology itself ... is able to do the work of social differentiation." (Mosse, 2008:941-944).

These words of Mosse (2008) capture the dynamic interactions between the natural and social realms of water through time and space. He argues that these interactions are ever changing and highly contextual, yet at the same time he shows that material artefacts such as hydraulic property mediate and even can solidify these interactions, inscribing them on the landscape. In her research on social differentiation, Nightingale (2011) shows that *"... social hierarchies become materially manifest on the landscape as people are involved in work practices that offer them possibilities to contest, resist or conform to subjectivities"* (Nightingale, 2011:154). Both authors conclude in their research that the social and the natural are closely intertwined. For many years scholars from different disciplines have therefore argued that views on society and nature as separate domains are artificial and problematic (Latour, 1993; Harvey, 1996; Swyngedouw, 1999; Castree and Braun, 2001; Budds, 2008; Mosse, 2008; Nightingale, 2011; Di Baldassarre et al., 2013). I share their view and consider the social relations and natural processes to concurrently constitute and reorder physical environments forming dynamic waterscapes. Building on political ecology theories, the waterscape can be understood as a produced socio-natural entity in which the environment is not regarded as simply *"a stage or arena in which struggles over resource access and control takes place ... [but where] nature, or biophysical processes, ... play an active role in shaping human-environmental dynamics"* (Zimmerer and Bassett, 2003:3). It simultaneously conceptualizes the physical hydrological cycle and biochemical processes as well as the ways in which water is controlled and shaped by social relations and institutions (Budds, 2008). In other words, the waterscape is thus not only produced by the habitation of nature through the damming of rivers, the diversion of the water flows and the construction of infrastructures to distribute water to users (Loftus, 2007) and the responses of nature to this occupancy (Budds and Sultana, 2013), but also by the agency of the physical environment itself. As the domination of humans over nature remains incomplete, ecological relations shape and reshape societies and circumscribe the ever changing range of choices available for human exploitation. The materiality of water, especially its variable, forceful, fugitive and directional flow, gives also a clear example of how agencies of non human natures can be a constitutive element of social processes. For instance, in the case study in Tanzania I will show how upstream-downstream asymmetries in availability of water in a catchment fosters intermarriages between smallholder farmers in an attempt to secure livelihoods and in the case study in South Africa I will discuss how the siltation of a downstream reservoir forces a commercial farmer to maintain a constructive relationships with an upstream community (see chapter 2 and chapters 3 respectively; see also Van der Zaag, 2007; Mul et al., 2009; Komakech et al., 2012b). Moreover, the agency of the physical environment is also exercised by (hydraulic) infrastructure. Not only does hydraulic infrastructure organize space, once constructed, it also becomes a force in itself, capable of rearranging and affecting water flows, often outliving the particular alliances who constructed it (Mosse, 2008). As such hydraulic infrastructure is not merely a passive instrument of human will, but an agent that actively opens certain trajectories while foreclosing other, alternative pathways in society (Swyngedouw, 1999; Ahlers et al., 2011; Meehan, 2014; Van der Zaag and Bolding, 2009). The waterscape is thus dialectically produced by actors of human and non-human nature in an ever ongoing process. Moore (2011) therefore concludes that *"the environment as a single object does not exist "because every species, not only the human species, is at every moment constructing and destroying the world it inhabits""* (Lewontin and Levins 1997: 98 as cited by Moore, 2011: 6).

In the constant reordering of environments, unequal social relations play a central role *"in determining how nature is transformed: who exploits resources, under which regimes and with what outcomes for both social fabrics and physical landscapes"* (Budds, 2008:60; see also Leach et al., 1999; O'Reilly et al., 2009). Waterscapes are thus never neutral but represent as well as shape dominant political regimes and as such are continuously contested by rival actors (Haraway, 1991; Swyngedouw, 1997; 1999; Zimmerer and Bassett, 2003; Budds, 2008; O'Reilly et al., 2009; Budds and Sultana, 2013). Analyzing waterscapes from this politicized perspective can help us to understand how mutually constitutive social-natural processes produce, propagate and reinforce particular water resource configurations. Or as Swyngedouw phrases it: *"the flow of water, in its material, symbolic, political, and discursive constructions, embodies and expresses exactly how the 'production of nature' is both arena for and outcome of the tumultuous reordering of socio-nature in ever changing and intricate manners"* (Swyngedouw, 1999:449). In her research on the production of social differences, Nightingale (2011) argues that: *"... the movement and actions of bodies in space produce power-laden boundaries between people that are open to contestation, but also partially close down possibilities for radical transformation of social inequalities ... with significant material consequences ..."* (Nightingale, 2011: 155). This emphasizes the materiality of uneven social relations and the intimate link between authority and access to physical resources. Social difference and uneven institutions are thus not only reproduced through normative orders that structure society, but also through processes of material accumulation and dispossession. This combined socio-nature approach has allowed political ecologists to study physical processes of differentiation and the associated transformation of waterscapes under capitalism (Harvey, 1996; 2003; Loftus, 2007; Budds, 2008; Bond, 2006) as well as the pervasiveness of neoliberal shifts in waterscapes across the globe despite the context specificities of how these shifts unfold at local level (Harvey, 1996; Haraway, 1991; Swyngedouw, 1997; 1999; Harris, 2009). Beyond critiquing the neoliberal project, I will use this approach to critically analyze how waterscapes are formed through everyday practices circumscribed by plural normative orders in society. Deconstructing the waterscapes thus helps me not only to understand how water resource configurations are embedded in the political and economic doctrines, but also within the cultural practices of the actors involved (Mosse, 2008). In this analysis, different yet interlinked spatial and temporal scales are crucial aspects, not only in view of political economic influences beyond the local level and the present time (Blaikie, 1985; Budds, 2008), but also because small changes in socio-nature relationships, even those occurring at extremely fine scales, can exert substantial influence on processes manifest at significant larger scales (Sneddon et al., 2002).

1.3 Objectives of the dissertation and research questions

Since the 1980s a major change took place in public policies for water resources management. This change in policy approach has been criticized based on empirical research on two main grounds; *one)* the policies resulting from this approach often do not achieve what they envision on paper due to interpretation, negotiation and rearrangement by socially positioned actors at local level leading to context specific outcomes; and *two)* the policy approach has paved the way for the proliferation and implementation of similar policy models in dissimilar contexts resulting in generic, decontextualized outcomes. These findings create a paradox in science: on the one hand the mainstream policies approach is perceived to have only limited influence on water resource configurations in waterscapes, yet at the same time it is argued it affects similar patterns in water configurations in different waterscapes around the world. To unravel this paradox, the overall research question in this dissertation is:

To what extent, how and why does the mainstream approach in water policy reforms influence, shape and change the water resource configurations within waterscapes?

To answer this question, this research examines the interplay between public policies designed and implemented by government agencies and the institutions that govern access to, control over and distribution of agricultural water resources. How this relationship unfolds within waterscapes that are historically constituted by natural and social processes is the overarching research objective of this interdisciplinary study. In addition, I have set two specific objectives to ensure the societal and scientific relevance of this research, namely:

First, to attempt to develop tangible directions for revisiting public policy processes within the water realm, in order to regain responsiveness to dynamic socio-nature processes and to address structural inequities.

Second, to attempt to enhance theories on critical institutionalism by explicitly linking them with theories that focus on constitutive spatial and material processes, and applying the theoretical frame to empirical case studies.

To answer the overall research question, I will attempt to find answers to the following specific research questions (see also Figure 1.1):

1. How do institutions related to water emerge, transform and endure and how do they constitute and govern water resource configurations?

2. How do material artefacts, physical processes and the materiality of resources mediate and constitute water resource configurations?

3. How are the policy models selected for the water reform processes interpreted, rearranged and used by actors within their discourses, strategies and negotiations over water?

4. Which normative orders are underlying institutions and policies around water resources and how do these normative orders shape the interplay between them?

5. How are patterns in water resource configurations maintained and/or modified within waterscapes as result of the interplay between existing institutions and water policy reforms?

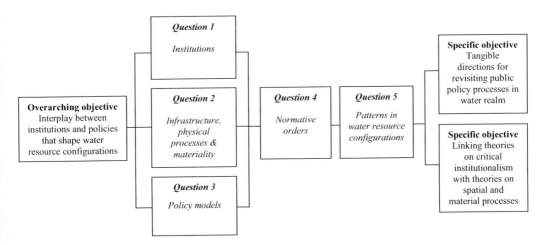

Figure 1.1: schematic overview of the research objectives and the linkages with the research question

1.4 Research methodology

1.4.1 Epistemological considerations

"Objectivity in research refers to doing justice to the object of study"
(Smaling, 1989: 307)

This research is based on an epistemological understanding that differs from the positivist scientific inquiry paradigm that is still dominant within the water domain and development orthodoxy. Instead of placing exclusive emphasis on the existence of logical relationships between evidence and truth claims, I embrace a more complex relationship between scientific knowledge claims and empirical evidence. I reject the view that objectivity in science can only be achieved based on conclusive evidence as this rarely, if ever, can be reached (Babbie and Mouton, 1998). Judgments by socially positioned scientists will therefore always affect scientific claims and understanding these inherently biased interpretations is crucial for scrutinizing knowledge claims (Brown, 1988; Limb and Dwyer, 2001). This even becomes more essential if we consider the constitutive relationship between power and knowledge; for information to be labeled as a 'fact' it needs to go through a process of ratification by those in position of authority and thus uneven relations of power and knowledge production are closely intertwined (Foucault, 1980; Mills, 2003). Objectivity in science therefore refers for me not to impartiality or conclusiveness, but to sets of procedures and methods used in science to obtain empirical evidence and build up arguments. What these objective procedures and methods can be is dependent on the subject of the research (Smaling, 1989), though it is the obligation of the scientists to make them explicit and transparent and justify why this approach is taken. On this basis the knowledge claims can be scrutinized by other scientists and complemented or contradicted by other empirical evidence or different reasoning.

Objectivity thus does not only come necessarily from repeating data sets, but by making the assumptions and choices made within the research clear and plausible to other scientists. This epistemological stance means that knowledge claims might be accepted by particular scientific communities, but will always remain inconclusive and possibly might be rejected in the future in the light of new empirical evidence and alternative arguments (Babbie and Mouton, 1998).

How this epistemological understanding is translated into research strategy, approach and methods is summarized in Table 1.1 and will be further discussed in the next paragraphs. With these methodological choices the research strives to do justice to the object of study and aims to produce reliable data and research findings that will be recognized as objective scientific inquiry (Smaling, 1989; Babbie and Mouton, 1998).

Table 1.1: notions of objectivity and procedures applied in research

Notion of objectivity in qualitative science	Description	Procedures employed in this research
Credibility	Compatibility between the constructed realities that exist in the minds of the interviewees and those that are attributed to them by the researcher	Residing in study catchments, triangulation within and between methods and sources, detailed interview narratives, feedback from interviewees on research findings
Transferability	Compatibility between what the researcher writes and what the reader receives	Thick and comprehensive case-study descriptions, insight in sampling technique, explicit and transparent methodological choices, clear positioning of researcher
Dependability	Similarity of research findings when research is done by other researcher with the same subjects and in same context	Engaging multiple researchers, supervision by senior academics, publishing in peer-reviewed journals
Confirmability	Compatibility between the data collected and the research findings obtained	Making interview outlines, schedules, interview narratives, field notes, using secondary data, data analysis records, process notes, coding of interviewees, referencing to interviews

1.4.2 Research strategy

For unravelling the complex interplay between institutions and public policies as well as the constitutive relations with water resource configurations, I need to simultaneously study what actors think, write, say and do as well as following where the water flows through the waterscape. This requires conducting interdisciplinary and in-depth research in 'real life' contexts and therefore I chose case study research as the overall strategy for this enquiry (Yin, 2003). Case study research allows for studying trajectories of change over time and focuses on process rather than outcomes. Moreover, through case study research detailed insights can be obtained from different perspectives which give room for analyzing complexity and ambiguities. Within this overall strategy I have made two deliberate choices. The first choice is to study multiple cases in different African countries to explore and capture of what is believed to be divergent but increasingly interconnected trajectories of change within the globalized political economy (Hart, 2006; see also Harris, 2009; O'Reilly et al., 2009; Swyngedouw, 2011). In this way I attempt to contribute to an incorporated comparison

(McMichael, 1990; 2000) and create *"connections across diverse but interrelated arenas of struggle"* (Hart, 2006:988). This choice is a direct response to the aim of this research to explicitly incorporate local-global connections into critical institutionalism in order to study the interplay between global structural forces and local historic particularities. Moreover, incorporated comparison may help to understand the production of gendered, class, racial and ethnic forms of differences as active constitutive forces driving different trajectories of change and forging strategic alliances (Hart, 2002; 2006; Nightingale, 2011).

The second deliberate choice I made is to adopt the <u>extended</u> case study method *"... in order to abstract the general from the unique, to move from micro to the macro and connect the present to the past in anticipation of the future, all by building on preexisting theory"* (Burawoy, 1998:5). In his methodology, Burawoy critiques conventional case study approaches for being inherently particular, a-historic and confined to small geographical spaces (Burawoy, 1991; 1998; Burawoy et al., 2000; Burawoy, 2009). He therefore suggests to extent the case study approach in five different dimensions. The first dimension relates to extending the research towards the researcher. Based on the understanding that knowledge is always partial and situated and that it is impossible for a researcher to attain a detached observer position, Burawoy (1991; 1998) encourages not detachment but rather active engagement and dialogue with the subject of the research as the way to obtain knowledge (see also Foucault, 1980; Scott, 1986; Mills, 2003; Nightingale, 2003; Zwarteveen, 2006). This extension requires a reflexive research approach in which the researcher explicitly positions herself in this world as well as in relation to the subject and context of the study and attempts to identify and articulate the partiality that her own limitations, history and standpoint bear on the research. The second dimension relates to the subject of the study. Where conventional approaches often focus on extraordinary phenomena or special events, Burawoy suggests focusing on everyday life situations in order to compare *"... similar phenomena with a view to explaining differences"* rather than comparing *"... unlike phenomena with a view to discovering similarities"* (Burawoy, 1991:280). This will push the research beyond the particularities of the case study and potentially generate broader claims and understandings (see also Hart, 2006). The third dimension is extension in time. This builds on understanding that history shapes the present and that we thus cannot understand a contemporary situation without knowing the past. This extension involves conducting historic studies of the case study area as well as its actors. The fourth dimension is extension in space. In contrast to conventional case study approaches that aim to derive meaning about macro environments from micro situations, Burawoy (1991; 1998; 2009) emphasizes the need to also look beyond the case study area and see what happens at larger spatial scales and how this possibly constitutes what happens at smaller spatial scales. In this he aims to reveal the *"macro foundations of a microsociology"* (Burawoy, 1991:280) and capture local-global connections and constitutive processes. The last dimension refers to extension of theory in which Burawoy (1991; 1998; 2009) argues to build on preexisting theories through successively reconstructing theories by explaining contradictions arising from empirical data rather than to constantly focusing attention on developing new theories. In this research I aim to build further on the critical theory tradition (Horkheimer 1982; Babbie and Mouton, 1998; Limb and Dwyer, 2001). In this tradition the ultimate aim of research is to lead to transformation and progressive change of what is perceived as a structurally unfair world. Critical here refers to the attempt to unravel social structures that have led to inequity and as such critiquing the foundations on which society is built. Beyond explaining why things are the way they are, research within the critical theory tradition attempt to show how society could also be and as such ultimately strive for political emancipation. As such, Horkheimer argued that a theory can be considered as critical insofar it seeks *"to liberate human beings from the circumstances that enslave them"* (Horkheimer 1982:244). The ways in which I aim to contribute to the critical theory tradition are specified in the objectives that I have defined for this research (see paragraph 1.3) and in the next paragraphs I further elaborate how I will engage with some of the research approaches within this tradition.

Table 1.2 summarized how I have applied the five extensions as suggested by Burawoy (1991; 1998; 2009) in various ways throughout my research.

Table 1.2: implementation of extended case study method in research

Extension	Application in research
Towards researcher	Active engagement with subjects during field research, making theoretical choices explicit, identifying shortcomings and biases in research (see epilogue of dissertation)
Subject of study	Focus on everyday interactions between actors, observation of mundane activities such as distributing, accessing and using water, focus on average hydrological fluctuations (rather than extreme events)
In time	Including historical narratives of the case study areas, capturing concise personal histories of interviewees, analyzing perspectives, patterns and relations over time
In space	Engaging with the global-local dynamics and studying constitutive spatial processes, incorporated comparison between case studies to illuminate what they have in common at different scales, world historical moments and/or constitutive processes
Theory	Focus on advancing the critical institutional theory by linking it with theories that illuminate constitutive spatial and material processes and by reconstructing critical institutionalism through empirical research

1.4.3   Research approach and methods

In order to fulfil the research objectives and align with the selected theoretical and methodological frameworks, I have chosen to study how contemporary water reform processes have unfolded and shaped water resource configurations in different waterscapes. For this purpose I have selected four African countries that have gone through extensive water reforms during the last two decades (1990-2010) that shared common foci on:

1) Revisiting the water right system and securing water allocation through time-bound conditional water use permits for private entities
2) Decentralization of water management responsibilities and increasing stakeholder involvement through establishment of water user platforms at different spatial levels
3) Economization of natural resources by enforcing payment for water through charging fees for water use

Based on pragmatic considerations related to access to secondary data and possibilities for extensive field research, I selected four countries, namely Kenya and Tanzania in eastern Africa and South Africa and Zimbabwe in southern Africa. The study catchments selected in each of these countries are listed in Table 1.3 and indicated on the map in Figure 1.2. The actual name and location of the study catchment in South Africa will not be revealed due to ongoing political sensitivities between various actors in the area.

Table 1.3: Overview of countries, river basins and case-study catchments

Country	River Basin	Study site
Kenya	Upper Ewaso Ngiro North River Basin	Likii catchment
South Africa	Thukela River Basin	*Undisclosed*
Tanzania	Pangani River Basin	Makanya catchment
Zimbabwe	Save River Basin	Nyanyadzi catchment

Figure 1.2: Map of Africa with overview of approximate locations of case study areas

My methodological choice to opt for multiple extended case studies has had considerable consequences for my research approach. Inherent to case study research is that the content of the case cannot be fully controlled by the researcher as activities or situations may come up that were not foreseen and the subject may also influence the line of enquiry (Yin, 2003). The research questions have guided the data collected in the cases, though not in every case study each research questions has been explicitly addressed. Instead, each case study had an exploratory character with a different focus depending the specific context of the case study area as well as the way in which the research unfolded during fieldwork (see Table 1.4). Moreover, in order to obtain comprehensive understanding based on the extensions defined by Burawoy (1991; 1998; 2009), the selected methodology required me to analyze the data in a chronological order per case rather than per research question. Nevertheless, to allow for comparative analysis between the cases, each research question has been addressed by more than one case study and the findings of all four case studies are brought together to answer the broader research questions 4 and 5. Through this approach I attempt to achieve the research objectives. It should however be noted that the choice for multiple study sites also meant less time was available per case study which affected the quantity of the empirical data collected per case and thus also the analytical depth of the individual cases. To partly compensate this limitation, this research was largely carried out under a larger collaborative research project[13] and/or builds further on intensive research carried out by other researchers in the same case study catchments, which allowed for complementary research findings to enrich this study and vice versa (see amongst others: Bolding, 2004; Kongo and Jewitt, 2006; Enfors and Gordon, 2007; Kosgei et al., 2007; Makuria et al., 2007; Mul et al., 2008a; Mul et al. 2010; Bossio et al., 2011; Komakech et al., 2012b; Méndez et al., *forthcoming*)[14].

Table 1.4: Link between case studies and research questions

Case study	Focus on:	Contributes to:
Kenya	RQ1: Infrastructure	RQ4: Normative orders
	RQ3: Policies	RQ5: Water resource configurations
South Africa	RQ1: Institutions	RQ4: Normative orders
	RQ2: Physical processes & materiality resources	RQ5: Water resource configurations
	RQ3: Policies	
Tanzania	RQ1: Institutions	RQ4: Normative orders
	RQ2: Physical processes & materiality resources	RQ5: Water resource configurations
Zimbabwe	RQ2: Infrastructure	RQ4: Normative orders
	RQ3: Policies	RQ5: Water resource configurations

Within the four extended case-studies, I have combined elements of different research approaches. The main approach I employed is based on critical ethnographic studies (Burawoy, 1998, 2009; Hart, 2006; Atkinson and Hammersley, 2007). Originating from cultural anthropology, ethnography refers to rich descriptions based on direct observations of how people see, hear, speak, think and act in a particular (often foreign) society (Babbie and Mouton, 1998). Nowadays ethnography is widely employed in other social studies including in the extended case-study method that *"applies reflexive science to ethnography ... building on preexisting theory"* (Burawoy, 1998:5). Critical ethnography builds on critical theory and focuses on unravelling implicit values expressed by actors and illuminating unacknowledged biases that may result from these implicit normative stances (Madison, 2005). In this way critical ethnography seeks to reveal ideology within action and aims to understand the behaviour of actors within a particular historic context. Where conventional ethnography describes 'what is', critical ethnography also aims to analyze the why and the 'what could be' in order to unravel implicit power relationships and to counter perceived inequalities. According to Hart (2006) critical ethnography thus *"offers the vantage points for generating new understandings by illuminating power-laden processes of constitution, connection and disconnection, along with slippages, openings and contradictions, and possibilities for alliance within and across different spatial scales"* (Hart, 2006:982). For constructing the critical ethnographies within the extended case studies, I mainly relied on primary data collected during field work conducted under this research, either by myself or by research assistants[15]. The main sources of data in this research are the water users residing in the study catchments, which are mainly small-scale farmers and in some cases large-scale commercial farmers. In addition, information was obtained from government officials involved in the water reform processes as well as representatives from NGOs and research organizations active in the study catchment. To collect the data several interview techniques were employed, including semi-structured interviews, oral histories, group discussions and

---

[13] Smallholder System Innovations in Integrated Watershed Management (SSI) programme (Bossio et al., 2011).
[14] Mul et al. (2010) and Méndez et al. (*forthcoming*) are included as Annex 1 and Annex 2 respectively.
[15] This research has resulted in four Master of Science theses of the following research assistants: Kwezi (2010); Méndez (2010); Munyao (2011) and Chinguno (2012).

informal conversations. The interviews addressed amongst others issues such as personal histories, livelihoods, institutions around access to and use of water and land resources, involvement in water management and decision making processes, past and current water resource configurations and the ongoing water (and land) reform processes. The interviewees among the water users were selected by a stratified random selection procedure (Babbie and Mouton, 1998) to guarantee geographical spread within the case-study area, to obtain information from various types of water users, and to ensure inclusion of voices from different political affiliation, age, race, class and gender groups. This sampling technique entailed that interviewees were randomly selected, though if certain kinds of water users, social groups or geographical areas were underrepresented, interviewees were purposively selected from these groups and/or locations. To a lesser extent the snowballing sampling technique (Babbie and Mouton, 1998) was employed for exploratory purposes at the start of each field work and where I came across unusual perspectives that I wanted to further investigate. In total 175 people were interviewed through semi-structured interviews carried within this research (see Table 1.5) with a considerable number of these interviewees being interviewed several times throughout the research duration, either formally or informally. In Kenya and Zimbabwe interviews were carried out in the native language of the interviewees by research assistants, while in Tanzania and South Africa interviews had to be (partly) conducted through translators. The data collected through interviews was captured in detailed coded narratives to ensure confidentiality of the interviewees.

Table 1.5: Overview of actors interviewed through formal semi-structure interviews including the ratio between male (m) and female (f)

Case study	Small-scale farmers (m/f)	Large-scale water users (m/f)	Government officials (m/f)	Others (m/f)	Total (m/f)
Kenya	33 (18/15)	4 (4/0)	5 (5/0)	2 (2/0)	**44 (29/15)**
South Africa	38 (22/16)	18 (18/0)	11 (7/4)	6 (4/2)	**73 (51/22)**
Tanzania	25 (13/12)	-	2 (2/0)	3 (2/1)	**30 (17/13)**
Zimbabwe	21 (15/6)	-	6 (4/2)	1 (0/1)	**28 (19/9)**
Total:	**117 (68/49)**	**22 (22/0)**[16]	**24 (18/6)**	**12 (8/4)**	**175 (116/59)**

The data of the interviews was complemented with data obtained through focus group discussions, informal conversations, field observations, attendance of meetings, data records and reports available about the case study catchments. At least five focus group discussions per case study site have been organized, which were mainly used to cross-check preliminary research findings with a randomly selected group or to focus on a particular (sensitive) issue such as conflicts over water allocation, discrimination or corruption with a purposively selected group (e.g. elderly, women, socially excluded, particular political affiliation). The everyday practice of actors in relation to water, including operating infrastructures, fetching water, cleaning canals and irrigating land, have been captured through observations of the visible undertakings of actors (Bourdieu, 1977; van der Zaag, 1992). Meetings of water user associations and river basin organizations were attended to obtain insight in the topics discussed as well as to document the behaviors and dynamic interactions between the participants. Data bases and reports of authorities were used to obtain data on water permits, ownership of infrastructures and landownership. Thematic analysis (Petty et al., 2012) was used to explore variations, similarities, patterns and relationships within the data collected for the critical ethnographies. I used thematic analysis amongst others to study how actors identify themselves and depict other actors around them, how they frame the problems they perceive in relation to water and land resources, how they perceive past and current water

---

[16] This group mainly consists of large-scale commercial farmers which in the case study areas are male dominated and as such no female interviewees could be selected.

resource configurations, how they use discourses and normative orders in their negotiations over access to and control over water resources, and how they perceive the reform processes. Responses from interviewees in relations to these themes were captured, compared and combined with other data sources to unravel the complex social realities of the actors. For the reliability of the research, triangulation is essential in the qualitative research paradigm (Babbie and Mouton, 1998). Therefore, this research used multiple methods and multiple sources of data to construct the critical ethnographies.

Discourse analysis was employed to study the discursive and rhetorical devices within the policy documents (published in English language) to guide the water reform processes in the case-study countries. In particular, elements of critical discourse analysis were used which moves the research approach beyond the analysis of the structure and content of a particular piece of text and systematically relates the text to the institutional structures that prevail in the society in which the text is produced and utilized (Fairclough, 1996; Weiss and Wodak, 2003; Fairclough et al., 2011). Within critical discourse analysis a discourse is considered as a constitutive process *"in the sense that it helps to sustain and reproduce the social status-quo, and in the sense that it contributes to transforming it"* (Fairclough et al., 2011: 358). Linguistic conceptualizations of the world articulate social configurations of difference and dominance and as such words used in discourses have a normative effect on society. Critical discourse analysis aims to generate insight into the way a discourse reproduces and/or resists normative understandings and inequalities in society to reveal how uneven social relations are enacted through discourses. In this research I employed the following specific elements of critical discourse analysis to study the water reform policies (Babbie and Mouton, 1998; Fairclough et al., 2011):

- Analyzing how the problems are framed and structured within the policy documents and identify the narratives used to justify the reform process
- Analyzing the objects of the policy documents (e.g. water allocation, water use, water redistribution)
- Analyzing the subjects of the policy documents by identifying the roles present in the document and reflecting on the agency of a specific role (e.g. subsistence farmer, emerging farmer, commercial farmer)
- Analyzing the mechanism the policy documents proposes (e.g. water licensing, participation, payment for water, environmental water flows) and the kind of policy model it legitimizes
- Analyzing the image of the world that the policy document articulates (e.g. productive versus non-productive water use, agricultural versus industrial use, public versus private, formal versus informal organizations) and how it addresses objections to its terminology (e.g. unauthorized use, illegal use, inefficient use)
- Reflecting on normative orders underlying the terms and concepts used in the policy documents (e.g. ideologies, moral stances, political choices)
- Relating the normative orders underlying the policy documents with the institutional structures prevalent in society and reflecting on how the policy documents reproduces and/or resists inequalities in society

The above research approaches were combined with general geographic analysis to study the current water resources configurations within the waterscapes. In this analysis I looked at demography patterns, hydrology, geology, land demarcation, land use, water use, livelihoods, agricultural practices and waterscape features including hydraulic infrastructures. The main sources of data included field observations, maps, aerial pictures, satellite images, databases, scientific publications and project reports. Without engaging in actual flow measurements, I have attempted to quantify water use where relevant based on design parameters of hydraulic infrastructures, water distribution records, field observations, crop yields and secondary data sources. This allowed me to take into account the material realities of the various actors in the case study catchments within my analysis.

As a final step within my research I brought the four extended case studies together in an incorporated comparative analysis. Acknowledging that societies are related in time and space, as they form part of larger world-historical processes, comparison of similar, interconnected, processes in dissimilar locations can attach meaning to those processes beyond the particular moment and place (McMichael, 1990; 2000). In this incorporated comparison I brought four experiences into relation to one another via analysis of water reform processes, understood as political constructs produced by a global policy network, across four geographical locations. This comparison is not a prescribed procedure in which I compared more or less similar cases based on fixed units of analysis. After all, the small-scale farmers in the case in Zimbabwe have other means and constraints when it comes to accessing water than the small-scale farmers in the Tanzanian case study and the commercial farmers in the South African case study have different personal histories than the commercial farmers in the case study in Kenya. The agricultural water users as unit of analysis in this research are thus place and time dependent. However, since the water reform processes in the case study countries share the same epistemic origin, these "... *process-instances are comparable because they are historically connected and mutually conditioning"* (McMichael, 2000: 671). Hence, rather than comparing agricultural water users across the case studies per se, I compared the production of social difference that materialized through, amongst others, the disparate access to hydraulic infrastructure, various forms of land tenure and the different kinds of water user rights, as result of the global shift in public policy approach within the water domain. The case studies were thus not considered as separate processes with common or contrasting patterns of variation, but rather as parts of a larger world-historical process in which this policy construct was produced and are brought together in order to understand this particular process by deconstructing how it affects water resource configurations in various waterscapes.

1.5 Structure of the dissertation

This dissertation consists of seven chapters. Chapter 1 discusses the societal and scientific relevance of this research and presents the main theoretical and methodological considerations. In chapters two to six the four comprehensive extended case studies are presented based on the empirical data collected under this research. As explained in section 1.4.3 not all case studies have a similar focus and they address different research questions. The order in which the case studies are best portrayed is therefore not straightforward. I have opted for the chronological order in which field data was collected as this order best shows the progressive insight that was developed during this research. The extended case studies have been published (or are in the review process for publication) by peer-reviewed journals[17].

---

[17] Since the case studies have been published in different academic journals and the research was conducted over a longer period of time (2007-2015), the writing style and terminology used in the chapters is not always consistent, reflecting the various inputs of co-authors, peer reviewers and editors as well as the vast growing literature base on which this research draws. Nevertheless, the main concepts used within this research are defined and discussed in this first chapter of the dissertation.

Chapter 2 discusses how institutions on water sharing emerge, transform and endure in a smallholder irrigation system in Tanzania. The chapter analyzes how these institutions are negotiated, agreed upon and contested based on various normative orders, questioning the existence of universal values of hydrosolidarity and placing the emerging institutional arrangements within the dynamic physical landscape (Kemerink et al., 2009).

Chapter 3 analyzes the normative orders underlying the water reform process in South Africa and shows how institutions prevailing in society shape the outcomes of the contested reform process and reproduces particular water resource configurations between subsistence and commercial farmers within a waterscape in which the actors are directly linked through the materiality of the resources (Kemerink et al., 2011).

Chapter 4 is also situated in South Africa, but specially focuses on how the policy model of decentralization through establishment of Water User Associations (WUAs) is interpreted, rearranged and used by actors within their discourses and strategies to legitimize access to and control over water at local level (Kemerink et al., 2013).

Chapter 5 discusses how existing hydraulic infrastructures intersect with policy models embedded in the water reform process within a waterscape in Kenya and analyzes how this leads to reinforcement of material differentiation between small-scale and large-scale irrigators (Kemerink et al., *forthcoming a*).

Chapter 6 analyzes the physical transformation of a waterscape in Zimbabwe in response to the water reform process and discusses how critical analysis of satellite images can be used to incorporate complex socio-nature processes into policy making process to aid policy makers who wish to respond to dynamic and context specific circumstances (Kemerink et al., *forthcoming b*).

The last chapter of this dissertation, Chapter 7, brings the extended case studies together in a relational analysis to unravel the interplay between policies and institutions as well as to discuss what the shift in policy approach has meant for the constitutive processes in waterscapes. In this way this chapter will reflect on the scientific paradox highlighted in chapter one and answer the overall research question. Moreover, this chapter seeks to achieve the specific objectives set for this research by discussing possible theoretical elaborations for critical institutionalism and offering bureaucrats concerned with equity issues tangible directions for revisiting the policy approach within the water realm.

In addition to these seven chapters, I have included two papers as annex to this dissertation, namely Mul et al. (2010) and Méndez et al. (*forthcoming*). These are papers that I co-authored and that provide a broader context of the Tanzanian and South African case studies presented in the dissertation and as such may be relevant for the interested reader.

2. Assessment of the potential for hydro-solidarity within plural legal conditions of traditional irrigation systems in northern Tanzania [18]

Abstract

Competition over water resources and related disputes over water are inherently local and context-specific in their manifestations. In Makanya catchment, located in the mid-reaches of the Pangani river basin in northern Tanzania, competition over water is apparent and with increased demands for water disputes are likely to become fiercer in the near future. Negotiations between upstream and downstream users at various levels in the catchment have resulted in water sharing arrangements or are still on-going while other negotiations seem to be stranded in impasses. Why in certain situations water sharing among users evolves, while in other cases mutual agreements cannot be reached, is not yet well understood. Insight in the plural legal context in which water sharing arrangement among water users develop could set light on complex resource use and management realities as well as the ability of various water users to influence the negotiations over water. The hydro-solidarity concept is referred to as potential mechanism to reconcile conflicts over water. Hydro-solidarity promotes ethical dimensions as integral part of decision making and is assumed to be based on a universal set of commonly accepted norms and rules. Through analysis of the plural legal context in which water sharing arrangements among the smallholder farmers in the Makanya catchment develop, the paper explores the existence of ethical dimensions in decision making and their legitimacy. In this way the potential of the hydro-solidarity concept as mechanism to reconcile conflicts over water can be assessed in the context-specific plural reality. The paper concludes that, although ethical dimensions in decision making in the Makanya catchment exist, the hydro-solidarity concept as mechanism to reconcile disputes over water has limited potential as long as it does not embrace the plural reality. The authors argue that, instead of searching for a universal normative order, legal plural analysis can serve as framework to identify the context-specific ethical dimensions in each of the normative orders that influence the negotiations over water. In this way hydro-solidarity can be strengthened from within each normative order respecting its legitimacy and acknowledging the water users using the normative order to claim their rights. Potentially this will lead to a more realistic mechanism to reconcile conflicts over water.

[18] This chapter is based on: Kemerink, J.S., R. Ahlers, P. van der Zaag (2009) Assessment of the potential for hydro-solidarity in plural legal condition of traditional irrigation systems in northern Tanzania. *Physics and Chemistry of the Earth* 34(13-16): 881-889

2.1 Introduction

Competition over water resources are not standard problems for which universally valid solutions can be formulated. Although partly attributed to the same causes, and although believed to be increasingly urgent in many places of the world, water problems are inherently local and context-specific in their manifestations. They are not simply reducible to natural and physical processes of water extraction and storage and do not follow universal economics or natural laws. Water control problems are both physical- ecological and human-made, the locally specific outcome of social and political histories and processes (Boelens et al., 2005).

In Makanya catchment, located in the mid-reaches of the Pangani river basin in northern Tanzania, competition over water is apparent and with increased demands for water (Grove, 1993; Potkanski and Adams, 1998) disputes are likely to become fiercer in the near future. Negotiations between upstream and downstream users at various levels in the catchment have resulted in water sharing arrangements or are still on-going while other negotiations seem to be stranded in impasses (Kemerink et al., 2007; Mul et al., 2010[19]). Why in certain situations water sharing among users evolves, while in other cases mutual agreements cannot be reached, is not yet well understood in the Makanya catchment. Insight in the legal context in which water sharing arrangements among water users (e.g. individuals, institutions, sectors) develop could provide understanding of the complex resource use and management realities as well as the ability of various water users to influence the negotiations over water. Analysis of the legal plural conditions can therefore be used as theoretical framework to assess the relevance of the conceptual mechanism of hydro-solidarity in which reconciliation of conflicts over water is based on solidarity among water users.

This paper describes the development of water sharing arrangements among the smallholder farmers in the Makanya catchment and analyzes the legal context in which the negotiations over water take place. The field research carried out in the catchment has built further on research conducted under the Smallholder System Innovations in Integrated Watershed Management (SSI) Programme (Bhatt et al., 2006). The research focused on the indigenous Manoo irrigation system in the mid-reaches of Makanya catchment currently serving a total of 134 smallholder farmers. The findings presented are based on in-depth semi-structured interviews with over 25 smallholder farmers within the Manoo irrigation system which were carried out between April and July 2007. The interviewed farmers were selected with a stratified random selection procedure to guarantee geographical spread, balance in age and gender and to include less advantaged farmers. The findings of the interviews were cross-checked through focus group discussions, observations, comparison with existing documentation of the case study area and by consultations of key-informants such as extension officers, local authorities and nongovernmental organisations (NGO) active in the region. This paper first gives some theoretical insights on hydro-solidarity and legal pluralism. After that the case study area of the Makanya catchment is introduced followed by a detailed narrative of the history of the Manoo irrigation system to get more insight in the context in which the case study situated. Thereafter the existing water sharing practices among the smallholder farmers at various levels are analyzed within the plural legal conditions. In the concluding chapter the impact of legal pluralism on the development of water sharing arrangements will be discussed and the authors will reflect on the potential for the hydro-solidarity concept as mechanism to reconcile disputes over water.

---

[19] Mul et al., 2010, is included as an annex to this dissertation.

2.2 Theoretical insights: hydro-solidarity and legal pluralism

The concept of hydro-solidarity has been developed in the late 1990s to counter the on-going resource security politics in which scarcity in resources were regarded as a threat to the (economic) development of nations. As alternative to the water securitization discourse (Wouters, 1999; Turton, 2002), trust building and cooperation among water users was promoted by donor organizations through focusing on public debate, dialogues and levelling of the playing field of all stakeholders (Weaver, 1995; Patrick et al., 2006). In this light the hydro-solidarity concept was introduced as the 'ethical basis for wise water governance' and put forward as desired mechanism to reconcile disputes over water (Lundqvist and Falkenmark, 1999). Falkenmark and Folke (2002) have defined hydro-solidarity as the reconciliation of conflicts of interest with a solidarity-based balancing of human livelihood interests which should be achieved against unavoidable environmental consequences. The hydro-solidarity concept promotes ethical dimensions to be an integral part of decision making. These ethical dimensions should be based upon a more or less universal set of commonly accepted norms and rules (Lundqvist and Falkenmark, 1999). The hydro-solidarity concept stresses the positive effects of cooperation rather than the negative effects of competition, and could potentially contribute to an enabling environment in which water resources can be managed in a sustainable manner. However, the definition of hydro-solidarity does not identify the norms and rules on which the ethical dimensions should be based. Moreover, it ignores the inherently local and context-specific manifestations of water sharing practices by assuming that a universal set of commonly accepted norms and rules exists. Consequently the hydro-solidarity concept remains a 'buzz-word' that has not yet been able to capture the complex realities of water sharing practices.

Legal pluralism refers to the existence and interaction of different normative orders in the same socio-political space that affect and control people's lives (Von Benda-Beckmann, 1997; Bentzon et al., 1998; Boelens et al., 2005). In legal pluralism the different normative orders may originate from various sources such as political ideologies, economic dogmas, religions and cultures, and are therefore in their very essence shaped by history and embedded in local realities. Hence, insights in plural legal conditions in societies can contribute to the understanding of complex resource use and management realities (Boelens et al., 2005). Not all normative frameworks have the same coercive means and power of enforcement with respect to the rules and regulations they produce and represent, nor do all enjoy the same degree of legitimacy and respect for their rules, rights and authority. Hence, analysis of the relationship between rights and power, as well as of perceptions and relations of legitimacy in legally plural situations is crucial as it defines the social meaning of rights (Merry 1992; Von Benda-Beckmann, 1997, 2002; Spiertz, 2000).

As the hydro-solidarity concept is supposedly based on universal set of commonly accepted norms and rules it assumes that this consensus exists or can be reached with the legal plural reality. Analysis of the various normative orders which influence the negotiation over water as well as their level of legitimacy could provide insight if this is viable. Hence, the analysis of the legal plural conditions could be used as framework to assess if hydro-solidarity can be embedded in reality. Based on the empirical data of the case study this paper will explore the impact of legal pluralism on the development of water sharing arrangements among smallholder farmers in the Makanya catchment. It will assess the existing water rights, the various power relations between the water users and the legitimacy of authority, which defines the social meaning of the rights and therefore influences the water allocation practices. In this way the paper will analyze the existence of ethical dimensions in decision making based on commonly accepted norms and rules that could indicate the potential for the hydro-solidarity concept in reconciliation of disputes over water.

2.3 Introduction to the case study area

The Manoo irrigation systems lies in the Makanya catchment (300 km^2) located in the South Pare Mountains, which forms part of the Pangani river basin in northern Tanzania (42,200 km^2, Figure 2.1). The Makanya catchment has a bi-modal rainfall pattern, receiving rainfall in two seasons per year with an average range of 400–600 mm/a (Makurira et al., 2007). Statistical analysis did not show significant changes in the total amount of rainfall over the past 50 years, however, it did show almost a doubling of the frequency of dry spells longer than 21 days during the long rainy season (Enfors and Gordon, 2007). The Makanya catchment is a semi-closed system in which the perennial rivers originate from the South Pare Mountains on the western side of the catchment, which rise to 2,100 m. Several tributaries join to form the main stream in the Makanya catchment. In the valley of the catchment the majority of the runoff is recharging the local aquifer under the sandy river bed (Mul et al., 2007). Nowadays only floods reach the outlet of the catchment, however, the river used to be perennial up to the late 1970s (SWMRG, 2003). In the South Pare Mountains the landscape is dominated by crop land on terraces, forest and bare steep mountains while in the midlands the landscape consists largely of cultivated land alongside bush land. In the lowlands the climate is considerably drier, with scattered low-growing bushes and solitary trees (e.g. baobab and acacia) characterizing the savannah-like landscape.

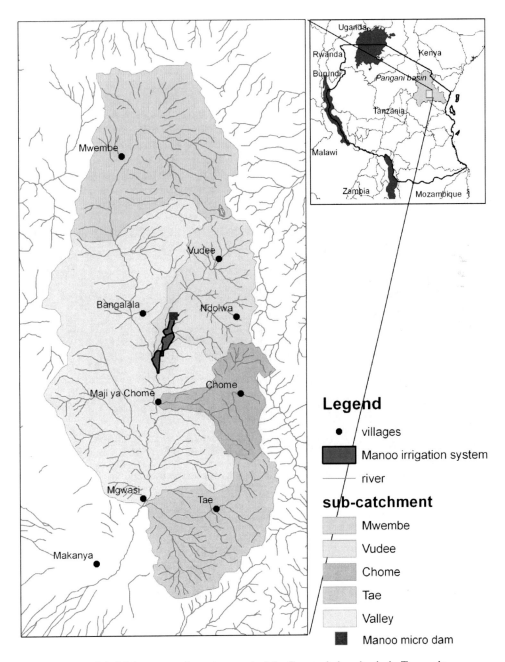

Figure 2.1: Makanya catchment as part of the Pangani river basin in Tanzania.

The Manoo irrigation system (Figure 2.2) is located in the middle reaches of the Makanya catchment with several other irrigation systems located upstream and an adjacent irrigation system with an intake on the other side of the river (Kemerink et al., 2007; Mul et al., 2010). The system takes water from the Vudee river downstream of the confluence with Ndolwa river. The main canal has a total length of approximately 3.5 km and crosses over two significant gullies. Undesired losses to lateral canals and natural drainage systems are minimized by closing off intake points with stones and earth bunds. Not far from the beginning of the intake is the Manoo micro-dam. The current capacity of the reservoir is 1,620 m^3 serving an area of circa 400 ha[20] (SAIPRO, 2004). However, the relative small size of the dam results in insignificant contribution of supplementary irrigation for dry spell mitigation mainly as a result of the large size of the command area and the large number of farmers participating in the irrigation system (Makurira et al., 2007). This analysis was confirmed by a study done on coping strategies during the dry spell period in 2005–2006 between users and non-users of the micro-dams, which showed insignificant differences of the impact of the dry spell on the livelihoods of the two groups (Enfors and Gordon, 2006)[21].

Figure 2.2: Manoo irrigation system, including the micro-dam and canal (Kemerink et al., 2007).

2.4 History of the Manoo irrigation system

The Makanya catchment is home of the Pare tribe, who settled in the region at least 16 generations back and are descents of several clans living in northern Tanzania and southern Kenya (Mshana, 1992). The Pare people were organized in different patrilineal[22] clans each headed by an elderly man. The clans were specialized in agriculture, blacksmithing, animal husbandry and worshipping, which translated into social classes among the clans. Although

---

[20] Maximum potential command area in rainy season, in dry season only about 6 ha is irrigated.

[21] Both groups adopted similar coping strategies (e.g. spending their savings and involvement in other economic activities than agriculture) and both were dependent on the natural environment and external aid for food during the dry spell.

[22] In patrilineal communities the heritage of property and power positions is through the male lineage, from father to son. Daughters marry outside the family, dowry is paid to the family of the daughter and children belong to the family of the father. In patrilineal Pare communities women have low social and political status and are not allowed to own resources such as land (Mshana, 1992; SAIPRO, 2004).

violent conflicts over natural resources between the clans have been documented the economic specialization fostered cooperation among the clans and semi-centralized chiefdoms of several clans were formed in the 17th century (Mshana, 1992). The chiefdom in the South Pare Mountains ruled a vast area that included the whole Makanya catchment. The chief had the main authority and was directly involved in natural resources management, including water resources. In the Pare culture it was believed that natural resources belonged to God and could not be owned by a person, but only be used by paying local brew to the chief (Mshana, 1992). In the past population density was low and the Pare lived up in the mountains where the best farming lands were located, away from the feared Masaai herding their cattle in the lower-lying areas and the wild animals (Mshana 1992; Kimambo, 1996). During the colonial era[23] cash crop cultivation (sisal, cotton and coffee) was imposed by the colonial administration through the chiefs who kept control over the territory (Sheridan, 2004).

According to the respondents the Manoo irrigation system was established in the early 20th century and started with a single canal. In 1936 the micro-dam of the system was built by about twenty families all belonging to the Wadee clan. Originally the dam was built to water cattle, but later on they started to store irrigation water as well. All clan members had several hectares of land in the current upstream part of the irrigation system on which they practiced rotational agriculture with supplementary irrigation. At that time farmers from other clans practiced rainfed agriculture and herded cattle downstream in the lower plains and once the increased yields obtained in the irrigation system became clear they requested access to the system. According to the farmers of Manoo extending the membership to families of other clans was (partly) driven by the need for cooperation amongst others to defend the village and irrigation system against attacks from the Masai tribe. It was the leader of the Wadee clan who decided who could become a member and normally local brewed beer was paid as membership fee. In the beginning families of other clans joined on the tail end of irrigation system, but after a while these families started to clear the bush further downstream and used the water from the system for irrigating the land. In this way the new members also obtained several hectares of land in the irrigation system and the command area of the system increased during the years. Although farmers from other clans could join the irrigation system, priority rights on the irrigation water were given to the Wadee clan members. One of such priority rights was that the clan members were given water during the day, while other people were given water during the night. Marriages between people of different prominent clans were viewed as a way to strengthen power positions in the irrigation system or were used to secure livelihoods by marrying people from other (upstream) irrigation systems and in this way obtain access to plots in different agricultural zones. With increasing population growth more farmers joined and rotational practices of agriculture were abandoned. The authority within the irrigation system was devolved through heritage following the traditional norms and customs of the Wadee clan.

After independence in 1961 the authority of chiefs over natural resources was abolished and the Manoo irrigation system came under the authority of the government of Bangalala village[24]. Under influence of the African socialism movement the irrigation system became formally the property of the whole community (Kimambo, 1996). After economic liberalization in Tanzania and introduction of the multiparty-system in 1995, democratic

---

[23] Tanzania was under German colonial rule (end of the 19th century until 1918) and under British colonial rule (1918–1961).
[24] Bangalala village has about 3300 inhabitants (2007) and is divided in 14 subvillages. Manoo irrigation system serves the three most downstream sub-villages of Bangalala. The village government consists of 25 elected (unpaid) representatives, including the chairperson and the representatives of each sub-village, and is supported by appointed (employed) extension officers specialised in e.g. agriculture, livestock, health and education.

governance structures were introduced in the irrigation systems and the management of the irrigation system was delegated to an elected water allocation committee. Between 2002 and 2003 the Manoo micro-dam was rehabilitated by a local NGO using a demand-driven and participatory approach[25], entering the community through the formal channels of the village government (SAIPRO, 2004). On advice of the NGO the constitution of the irrigation system was written down and families had to register themselves as members of the system. The members also agreed to pay a membership fee and seasonal fees for water allocations.

Most of the current 134 members[26] inherited the land from their parents. Only a few farmers bought their land, and some of them have plots in different zones of the Manoo irrigation system or even different irrigation systems. Maize and beans are grown as staple food mainly in the more downstream flat and fertile areas of the irrigation system, while onions, tomatoes, green peppers and cabbage are mainly grown on terraces in the steeper upstream parts of the system. Not all land within the irrigation system is under permanent cultivation. Relatively large parts consist of bush, which is used for grazing cattle. The water from the irrigation system is also used for domestic use (e.g. washing and bathing) and watering the cattle at the homesteads, especially because the piped water for domestic use is not very reliable in the part of the village where Manoo is located.

2.5 Impact of legal pluralism on water sharing practices

In line with the Pare culture the interviewed farmers still believe that natural resources are sacred and cannot be owned by human beings. Hence, most respondents indicate that all people living in the catchment should have equal access to water and conflicts over water should be solved through mutual agreements. However, currently the Manoo irrigation system has only an agreement with the adjacent Mkanyeni irrigation system on abstraction rotation and not with the upstream irrigation systems. Moreover, although an elected water allocation committee is in place, persistent clashes over water allocations between farmers within the Manoo irrigation system exist. This section will describe the existing water rights, the various power relations between the water users, the legitimacy of the different authorities and the actual water allocation practices. For practical reasons the case study is presented from the larger catchment scale to the smaller scale of the irrigation zones[27], however, this should not be interpreted as a causal relation in the system neither as the sequence of development of the system. It is interesting to note that, at larger spatial scale than discussed in this paper, water sharing agreements do exist between Bangalala and other villages within the Vudee catchment. Adherence to these agreements is confirmed by hydrological data (see Mul et al., 2010, in Annex 1).

---

[25] The NGO requested organized groups to define project ideas ranging from dam rehabilitation and livestock management to women empowerment projects and HIV/ Aids prevention. For the implementation of the projects the groups had to contribute 25% of the total costs of which 5% was in cash.

[26] Members registered in July 2007. This number varies per season as people who borrow land also have to register themselves and pay membership and water allocation fees.

[27] For management purposes, the Manoo irrigation system is divided into three zones: an upstream, a midstream and a downstream zone (Mul et al., 2010).

2.5.1 Water sharing with other irrigation systems

Since more than five decades the farmers of Manoo irrigation system have been discussing water sharing arrangements with the upstream irrigation systems. In some cases with success, however, the agreements were often abandoned during severe dry spells. Currently no water sharing arrangement exists even though the farmers of Manoo would desire to have an agreement. The upstream irrigation systems permanently abstract water without considering the water availability for the downstream irrigation systems (Kemerink et al., 2007; Mul et al., 2010). In line with the Pare culture, traditionally elderly men have been the main actors involved in negotiating water sharing among the different irrigation systems in the catchment. After independence the village government became formally responsible for the water supply to all residents in the village and water sharing among the villages became the responsibility of the districts. However, although formally elderly men are no longer responsible for water distribution between the irrigation systems, culturally they are still highly respected and therefore still actively involved in the negotiation over water. The interviewed farmers indicated that they feel the village government is too weak to dominate the traditional leaders, because they regard them as too young and inexperienced in farming to lead the negotiations over water. On the other hand, some Manoo farmers indicated that members of the village government are not willing to intervene, because they own plots in the upstream irrigation systems or have close relatives farming in the upstream systems. As a result of the limited involvement of the village government in the irrigation system, it is currently the Manoo water allocation committee together with the elderly advisors who are discussing the water sharing with upstream irrigation systems.

With the disadvantaged position of being located downstream of the Manoo irrigation system is dependent on the solidarity of the upstream users for their access to irrigation water[28]. One of the problems is the lack of reliable data on water availability in the different irrigation systems to support the claim of the Manoo irrigation system, which makes it easy for the upstream irrigations systems to deny that the water distribution is unfair. What further complicates the situation is that some farmers of Manoo own land or have access to land through family ties in the upstream irrigations systems and therefore do not acknowledge the dissatisfaction of the other farmers of Manoo. Respondents indicate that these farmers even prefer the current situation as they would like to keep their superior position and hence exert more power in the Manoo irrigation system. On the other hand the successful water sharing among the adjacent irrigation systems of Manoo and Mkanyeni is believed to be based on strong family ties between the farmers in the systems and therefore farmers have (access to) plots in both systems. This could mean that joint membership could also serve as an enabling factor, but potentially only if authority is rather balanced between the stakeholders like in the case of adjacent irrigation systems in contrast to upstream and downstream irrigation systems.

In the mean time the Manoo farmers have taken the modern 'democratic' governance system quite literally: one of the main reasons farmers indicate in the interviews for the increase of members within the Manoo irrigation system is the belief that joint forces will increase their chances to get access to upstream and/or alternative water resources. With more members[29] supporting this claim means more power in the democratic system of the village government.

---

[28] Hydrological data shows that a considerable amount of water entering the Manoo irrigation system originates from the same river as the upstream irrigation systems, probably from leakage of the upstream irrigation systems (see Mul et al., 2010, in Annex 1).
[29] Manoo irrigation system has together with the adjacent irrigation system approximately 310 members, while the irrigation systems directly upstream have ca. 240 members (see Mul et al., 2010, in Annex 1).

That this increase of members also has lead to more struggles over water within the Manoo irrigation system is by many farmers taken for granted. However, so far the democratic system is still contested by the other normative orders in the legal plural setting. The traditional authority of elderly people remains socially recognized and the Manoo farmers are socially connected to upstream irrigation systems through family structures, which influence the negotiations over water. In addition the executive capacity of the village government is limited to mediating on the water distribution to the irrigation systems and its functioning is adversely affected by party politics. Hence, the democratic governance system has not (yet) acquired enough legitimacy to result in a favourable water sharing arrangement with the upstream irrigation system.

2.5.2 Water sharing within Manoo irrigation system

Since the establishment of the system in the early 20th century, elderly men of the influential families of the Wadee clan have been the main actors involved in the water allocation to farmers in the Manoo irrigation system. After independence the village government became officially responsible for the water services in the village including irrigation water. However, priority rights[30] for clan members and succession of leadership within the system through heritage persisted until 1995. Thereafter, although with resistance from some of the members of the Wadee clan[31], the village government supported by NGOs introduced democratic elections of the water allocation committee and delegated the responsibility for water distribution between the farmers to the committee.

The water allocation committee consists of 10 members, including a chairperson, a vice-chairperson, a secretary, a treasurer, the three representatives of the irrigation zones[32] and one additional elderly advisor per irrigation zone. The committee meets once a week to discuss water allocations to three zones and other issues such as cropping patterns or communal work. The micro-dam of the Manoo irrigation system fills up three nights per week plus an additional night every two weeks altering with the adjacent irrigation system (Kemerink et al., 2007). Normally each irrigation zone is provided with water one day per week with the bi-weekly additional day rotating between the zones depending on which zone experiences the highest water stress, which is 'measured' by the requests for water from the farmers and not by physical observations of the crops. Transmission losses in the system[33] or number of farmers registered in the zones are not considered in the water allocation. During dry spells and the dry season the river flow is low and the Manoo dam does not fill up during the night. At those times water is only allocated to farmers in the upstream irrigation zone and the upstream farmers are requested to provide land to the downstream farmers. Sharing of land is in line with the Pare culture (Mshana, 1992), however, downstream farmers indicate that having family connections or close friendships with people upstream in the system is clearly an asset to ensure access to land upstream. Although considered as intrinsic part of Pare culture, it should be noted that sharing of the land is not institutionalized in the constitution of the irrigation system. Most downstream farmers give presents and part of the harvest as compensation for borrowing land from upstream farmers, although many downstream farmers

---

[30] See section 2.4 of this chapter.
[31] The Wadee clan member who was the leader of the irrigation system the moment the election was introduced refused to be candidate for the position of chairperson and some other clan members also boycotted the first elections.
[32] For management purposes Manoo irrigation system is divided into three zones, namely Kwanyungu (upstream, 40 members), Heiziga (midstream, 50 members) and Heishitu (downstream, 44 members).
[33] Transmission losses are estimated up to 80% for the downstream plots (Makurira et al., 2007).

indicate they do not consider this as payment, while downstream users are not compensated for not receiving the water nor is the seasonal irrigation fee returned to them.

Elections for the members of the water allocation committee take place every 3 years and are overseen by a representative of the village government. Close observation of the election process and analysis of the outcomes of the elections show that membership to the committee is kept within an inner-circle of farmers. The vast majority of the current members of the water allocation committee have held their positions for several terms and/or their direct family members have had a position in the committee as well. To elect the chairperson the water allocation committee nominates three candidates from which one is elected by the farmers. Prior nomination of candidates potentially leads to a stronger influence of the people represented in the committee. For example, in the last election held in May 2007 two female and one male[34] farmers were nominated by the committee as candidates for the position of chairperson, even though some farmers objected to their candidatures. Within the Pare culture the role of a leader of the group (e.g. chairperson) is still regarded as male position (SAIPRO, 2004), hence, it was a foregone conclusion before the elections that the new chairperson would be a male candidate.

Tension still exists between some of the original Wadee clan members and the other farmers in the system. As most clan members have considerable big pieces land in the upstream zone, they maintain an advantaged position in the irrigation system as they are closer to the dam. They claim their families should keep priority rights over water allocations and should not be obligated to pay the water fees. At least one of the Wadee clan members still refuses to pay the water allocation fees and data gathered from interviews and observations shows that his household receives considerable water. Although other farmers do not agree with the exception, they accept it to avoid conflict with the clan member. However, the power position of the Wadee clan members has been reduced over the years with the introduction of democratic ways of governance of the irrigation system. Nowadays there are few well-connected and wealthy families[35] downstream who seem to keep the power within their network through the election process of the water allocation committee as described above.

The difference in cropping patterns within the irrigation system due to the bio-physical differences between the cultivated zones (e.g. soil type and slopes) also influences the leverage positions of the farmers. Upstream in the system mainly vegetables are grown on terraces, while downstream on the more fertile and flat plains maize is grown. Trade in agricultural products between the zones within the irrigation system takes place and, as maize is the main staple food for the people, this economic specialization gives an advantage position to the farmers in the downstream part of the system. Moreover, farmers who own land in more than one zone benefit from this difference in cropping patterns. Both upstream and downstream farmers indicate that inadequate water in the tail end of the system affects the access to staple food in the area. However, the farmers also point out that direct trade was strong in the past, but this has weakened due to improved access to regional markets and an influx of imported products (e.g. cheap rice from India).

---

[34] The male candidate has been representative of Heiziga zone for over 6 years and he belongs to the Wadee clan. However, his father only became chairperson of the Manoo irrigation system after the introduction of the democratic elections.

[35] These families are often descendents from the less influential Wadee clan members and/or descendents of 'first generation' of families who joined the irrigation system after the irrigation system was established by the Wadee clan.

As described above the centre of power in the Manoo irrigation system has shifted and negotiations over water allocations are now taking place in a governance structure influenced by an externally introduced democratic process. However, the cultural beliefs and the traditional governance structures still have authority and family ties among farmers affect the democratic election process of the water allocation committee. On the other hand authority based on social grounds is partly replaced by economic power and bio-physical conditions offer opportunities to both upstream and downstream farmers. Over time the plurality in which the negotiations over water take place in the Manoo irrigation system has increased.

As result of this plurality the legitimacy of authority within the irrigation system has broadened and power is no longer in the hands of a few farmers belonging to the old Wadee clan. Other groups of farmers have successfully managed to claim their rights based on the various normative orders within the legal plural setting. However, the water sharing practice still represents unequal access to resources for the downstream farmers, especially during the dry period, as well as the less-connected farmers.

2.5.3   Water sharing practices at irrigation zone level

Water allocation to individual farmers is coordinated by the elected representative of the irrigation zones. The representative is supported by an appointed elderly advisor who is mainly involved in conflict mediation. Directly after the meeting of the water allocation committee the farmers can put in their requests for water to the representative of the zone in which they are registered. By allocating the water to the farmers the representative takes into account if the farmer already received water and if the canals and plots are well prepared to receive the water. Farmers who do not attend the water allocation meeting or are not represented by a neighbour will not get any water. Normally up to four beneficiaries receive water per day depending on storage available in the Manoo micro-dam. Usually there are no requests for irrigation when rainfall is adequate, but a week or more after a dry spell has started when crops start experiencing water stress many requests are received at the same time (Mul et al., 2010). On average farmers get one to two official allocations per season depending on the rainfall. During times of low rainfall farmers only get water to irrigate part of their land or for a shorter time frame, so that many farmers can be served in one day. Farmers receiving their irrigation turn are responsible for the distribution among themselves and for opening up the bunds. The water is spread on the fields using flood irrigation.

Farmers indicated that fierce competition takes place for the position of a representative of a zone, so that power can be kept within circles of family and friends. According to interviewed farmers people tend to support the person who might favour them with extra water allocations or water allocations on the right moment when crops start experiencing water stress. Several cases are known of representatives who have been fired by the chairperson and the elderly advisors of the water allocation committee on charges of corruption. This corruption entailed gifts of alcohol or sugar, which were traditionally accepted by the chief. Cases are also known of sexual harassment of female farmers requesting water turns by some representatives of the zones. One of the main problems to fight the above bribery is that no proper records[36] are kept on who got a water allocation, hence nobody can prove if the distribution of water has been fair or not. Farmers indicate that trying to introduce such an accounting system has led to opposition by people who are favored by the current system. Another problem is that the

---

[36] As indicated by the farmers the records should include the size of irrigated land, the duration of turn, the water availability during the irrigation, timing of the turn in relation to the growing cycle of the crops and should be signed for agreement by the farmer receiving the turn.

constitution does not specify on how water should be allocated and what the rights are of the individual farmers, which makes it difficult for a farmer to contest the distribution of water[37]. In case a farmer does not agree with the water allocation, he/she can complain to the representative of the zone. If they cannot come to an agreement they can ask the chairperson and the elderly advisors of the irrigation system to intervene. In case the chairperson cannot solve the problem the village government can be involved to mediate between the parties. However, the involvement of the village government is always through the water allocation committee and not through an individual farmer. Hence, individual farmers have limited possibilities to oppose decisions made by the water allocation committee unless their leverage position is as such that they can utilize other (informal) institutional networks. Revision of the constitution has been on the agenda of the water allocation committee for some years, but so far it has not been acted upon. According to the farmers this is partly caused by the unwillingness of the well-established farmers to adjust the constitution e.g. mainly the relatives of the representatives of the zones, but also influential members of the Wadee clan and upstream farmers. Another reason given by the farmers is that they need external support to revise the constitution. This high dependency on external support is a result of the approach adopted by NGOs as well as the legacy of colonialism.

Based on the advice of government and NGOs, farmers are aiming for gender balance in the water allocation committee. However, gender-inequalities still exist in the decision making within the irrigation system. Only a few women have become representative of their zone, but more often they have been elected to the position of secretary or treasurer in the committee. Most females indicate that the main reason why they do not want to become a representative of their zone is that the position of representative is highly demanding and people can get aggressive if they do not get water they have requested. Another explanation given is that traditionally the Pare tribe believed that the water source would run dry if a woman comes close to the resource during her menstrual period (Mshana, 1992). These kinds of beliefs potentially make it harder for women to distribute the water as, for instance, opening the water tap of the dam by women is still a taboo (SAIPRO, 2004). One of the main obstacles for gender equality is that women culturally still cannot inherit resources (e.g. land, water, cattle) from their parents. Women have access to resources of their male family members, but they do not have full control over the resources as their authority is limited within the official decision making structures. Buying of land by women is legally possible nowadays, but it is still very uncommon and considered as a taboo. Some female headed families (e.g. unmarried women and widows) indicate that they feel threatened and often resources (including water) and property are taken by male (in-law) family members. However, most other female respondents indicate that they do not feel discriminated as the gender-inequalities are in line with their culture and therefore are socially accepted norms. This structural inequality based on gender has been accounted in similar small-scale irrigation systems world-wide (Cleaver and Elson, 1995; Udas and Zwarteveen, 2005; Ahlers and Zwarteveen, 2009; Vera-Delgado and Zwarteveen, 2007).

Besides water allocations the committee is also responsible for appointing farmers to attend training courses offered by NGOs active in the Makanya catchment. Although these NGOs aim for equal access to knowledge, it was observed that some well-connected farmers

---

[37] The constitution only includes general articles on objectives of the group, membership characteristics, meetings to be held, the composition water allocation committee (including their general tasks) and a short article on conflict mitigation. It does not define how water is divided between zones and/or farmers, what the right and orders of the farmers are, nor does it mention which fees will be charged and where the fees will be used for.

attended several training courses, while others did not. Also the intended follow-up to train fellow farmers did not take place in many cases. Moreover, NGOs provide equipment and materials to individuals amongst others to plough the land, to spray pesticides, to construct terraces and to build rainwater harvesting tanks. Although it is demanded that the equipment is available for the whole community, it is often kept as private property. In this way access to NGO support becomes a source of power and hence NGOs play a significant role in influencing the leverage position of the water users. Furthermore, the advocacy of moral values by the NGOs (e.g. gender balance and democracy) strengthens the authority of specific groups of farmers which are not always the original target group (e.g. women and marginalized farmers). The NGOs indicate that limited financial and human resources force them to adopt simple approaches neglecting the power struggles among the farmers in the irrigation systems.

The analysis above has shown that the water allocation to individual farmers is controlled by the representatives of the zones. The legitimacy of their authority is given through democratic elections. However, the strong lobby during the election process indicates that social networks are an important mechanism to get access to water for irrigation. Normative orders such as the traditional view on family structures (clanship), cultural beliefs and traditions, and economic values can be identified as layers in the legal plural setting at this level. External interference in the irrigation system not only adds additional layers to the plural reality, but also directly influences the level of legitimacy of the various normative orders in which the negotiations over water take place. Within the current water sharing practice the less-connected and historically disadvantaged farmers are still marginalized as it is difficult for them to claim their rights within the legal plurality.

2.6 Discussion and conclusions

Within the Pare culture natural resources are sacred and cannot be owned by human beings. Therefore traditionally it is believed that people living in the catchment should have equal access to water and conflicts over water should be solved through mutual agreements. This implies an ethical base for the development of water sharing arrangements. However, often other behaviour is observed in the water sharing practice. In the past it was the chief who distributed natural resources between the people and priority rights for certain groups were common. Nowadays water distribution in the case study area is the responsibility of democratically elected village governments and under their authority water allocation committees have been established. Nevertheless, as described in this paper, not all negotiation processes have (yet) led to mutual agreements nor do all agreements represent equal access to water for the different (groups of) water users. As such it can be concluded that hydro-solidarity might play a role in the ideological view on water resources management among the smallholder farmers, however, it has little influence on the actual water sharing within the plural legality.

As described in this paper various normative orders influence the negotiation over water in the Makanya catchment. The legal conditions have become more plural over time as additional layers of normative orders have added to the legal spectrum. This has resulted in a current governance structure that is influenced by historical eras such as African traditionalism, colonialism, socialism, democracy and capitalism. This plurality has created a balance in power as the legitimacy of authority of the clans has been reduced and the authority of other groups has gained legitimacy. However, the legal plural conditions have not made the prospect of access to and control over water equal for all water users. Water users actively utilize the various normative orders that legitimize their claims and therefore serve their interests, so-called forum-shopping (von Benda-Beckmann, 1981; Meinzen-Dick and Pradhan, 2005). At the same time it is the leverage position of the water users that defines the extent to which they can influence the social meaning of rights and therefore the legitimacy given by the community to the various normative orders. Advantaged (groups of) water users utilize their authority and influence the outcomes of the negotiations over water. The paper illustrates that the level in which water users are advantaged or disadvantaged is shaped by historical markers (e.g. tradition and beliefs), bio-physical conditions of the farming land (e.g. location and soil) and socio-economic background of the water user. One important factor in the socio-economic background is the level of connectivity to other water users as claims on water are exerted through the social networks of the water users.

From the analysis above it can be concluded that ethical dimensions in decision making based on commonly accepted norms and rules do exist and potentially can be strengthened. However, in the complex reality of the legal plural conditions it is unlikely that they will gain sufficient legitimacy to become the dominant socially accepted norms if introduced as universal normative order. Asymmetrical distribution of power is an intrinsic part of every society and well-established water users will actively resist changes of water distribution except if it is an obvious win–win situation. Unless this is recognized within the hydro-solidarity concept, it risks staying a theoretical concept or at most becoming (yet) another normative layer added to the legal plurality in which water resources are managed. Therefore we argue that, instead of searching for a universal normative order, the hydro-solidarity concept should embrace the plural legal reality by incorporating the dynamic and context-specific conditions of water use. Legal plural analysis can serve as framework to identify the context-specific ethical dimensions in each of the normative orders that influence the negotiations over water. These dimensions can be different for the various normative orders. In this way hydro-solidarity can be strengthened from within each normative order respecting its legitimacy and acknowledging the water users using the normative order to claim their rights. Potentially this will lead to a more realistic mechanism to reconcile conflicts over water.

3. Contested water rights in post-apartheid South Africa: The struggle for water at catchment level [38]

Abstract

The National Water Act (1998) of South Africa provides strong tools to redress inequities inherited from the past. However, a decade after the introduction of the Act, access to water is still skewed along racial lines. This paper analyses the various ways in which the Water Act is contested, based on empirical data detailing the interactions between smallholder farmers and commercial farmers in a case study catchment in KwaZulu-Natal Province. The paper argues that the legacy of the apartheid era still dominates the current political and economical reality and shows how the redistribution of water resources is contested by the elite. The paper identifies several issues that prevent the smallholder farmers from claiming their rights, including the institutional arrangements in former homelands, the 'community approach' of Government and NGOs, the disconnect between land and water reform processes, and historically-entrenched forms of behaviour of the various actors. The paper concludes that the difficulties encountered in the water reform process are illustrative for what is happening in the society at large and raises the question as to what price is being paid to maintain the current status quo in the division of wealth?

---

[38] This chapter is based on: Kemerink, J.S., R. Ahlers, P. van der Zaag (2011) Contested water right in post-apartheid South-Africa: the struggle for water at catchment level. *Water SA*, 37(4): 585-594.

3.1 Introduction

Water use and management practices are a result of ongoing processes of negotiation and bargaining between different water users. The forum in which the negotiations over water take place is often characterised by legal pluralism, in which different normative orders coexist and interact (Von Benda-Beckmann, 1997; Bentzon et al., 1998; Boelens et al., 2005). Not all normative orders have the same coercive means; nor do all enjoy the same degree of legitimacy (Von Benda-Beckmann, 1997; Spiertz, 2000; Von Benda-Beckmann, 2002). At any location and in any point in time existing repertoires of water law and actual water use are therefore expressions of social-political and economic power relationships between people. Hence, proposed changes in water laws through water reform processes will often entail shifts in the socio-economic relationships which will benefit certain groups in society over others. Reform processes will therefore most likely be contested by some groups in society while other groups struggle to achieve the reformation (Mollinga, 2008; Mosse, 2008; Swatuk, 2008).

The impact of these power dynamics in society on the implementation of the water reform process becomes apparent in South Africa. The country has been haunted for decades by racial segregation under the so-called apartheid regime. The National Water Act (1998) formulated during the transition to the post-apartheid era is widely recognised in policy circles as one of the most comprehensive water laws in the world (Merrey, 2008). The Act defines the state as the custodian of the nation's water resources and only water required to meet basic human needs and maintain environmental sustainability is guaranteed as a right (RSA, 1998a). This fundamentally moves away from the previous water acts which were largely based on riparian water rights. The new Water Act gives the state a strong tool to redress race and gender inequities inherited from the past (Van Koppen and Jha, 2005). However, the National Water Act is implemented and enforced in a society thick with historically-entrenched socio-economic and political inequities. Hence, a decade after the introduction of the National Water Act access to water is still highly stratified along racial lines (Bond, 2006; Merrey, 2008; Cullis and Van Koppen, 2009). Recognition and understanding of the challenges met in the reform process could shed light on the forces at play in the resistance to redistribute the water. These insights can potentially contribute to better comprehending the struggles in the society at large.

This paper contributes to the ongoing discussion on the implementation of the South African National Water Act by presenting empirical evidence and analysing how the Water Act's goal to redress inequity is implemented and contested. This is done by illustrating the challenges faced by smallholder farmers in their struggle to increase access to (productive) water sources and their interactions with the commercial farmers in the area. The catchment used as a case study for this paper is located in the Thukela River basin in the south-eastern part of South Africa. The actual name and location of the catchment will not be revealed due to ongoing political sensitivities between various actors in the case study area. (References used in this paper that directly refer to the case study area will be indicated with 'undisclosed reference'. For verification purposes the references can be requested from the authors.) The findings presented are based on in-depth semi-structured interviews with twenty smallholder and commercial farmers within the catchment, carried out between June and August 2008. The interviewed smallholder farmers were selected by a stratified random selection procedure to guarantee geographical spread, to create a balance of age and gender, and to include less-advantaged farmers (see Box 3.1). The findings of the interviews were cross-checked through focus group discussions, observations, comparison with existing literature and by consultation of informants such as local authorities and non-governmental organisations (NGOs) active in the region. This paper first describes the theoretical framework used to analyse four ways in which water laws can be contested. In the next section a detailed narrative of the catchment is provided as well as the historical and institutional context in which the case study is situated. Thereafter the interactions between the actors in the catchment and the way in which they exert their water rights are analysed based on the theoretical framework. In the concluding section the impact of the contested Water Act on the access to (productive) water is discussed and the authors reflect on the consequences of the current situation for the society at large.

Box 3.1: Description of sample

The 18 selected smallholder farmers for the in-depth interviews permanently reside in the catchment in contrast to other residents who regularly commute to urban areas for longer periods of time. They are members of households located in different parts of the community and include members of large extended families residing in the area as well as farmers with few relatives living in the catchment. They have different political affiliations, sources and levels of incomes and educational backgrounds. All respondents are involved in agricultural activities and most own cattle. Out of the 18 smallholder farmers interviewed, 8 were women, of which 50% were older than 50 years and 50% were born in the community. Of the interviewed men, 70% were older than 50 years and 70% were born in the community.

The commercial farmers interviewed own the farms located directly downstream of the smallholder farmer community, including the commercial farmer whose property is (partly) located within the catchment. They are both male, born in South Africa from British descent and under 50 years old. The commercial farmers live with their families on the farm.

3.2 Theoretical framework: contested water rights

The main theoretical framework adopted in this paper is the concept of legal pluralism, which refers to the coexistence of and interaction between various normative orders in a society that govern people's lives (Von Benda-Beckmann, 1997; Boelens et al., 2005). The normative orders originate from different sources, such as political ideologies, economic doctrines, religions and projects. Legal pluralism recognises the dynamic, hybrid and ambivalent forms of laws that result from interaction between these normative orders within society, and allows for a greater understanding of the actual social meaning of rights in a specific social context (Boelens et al., 2005). Water sector reforms often aim at changing socio-economic relationships between water users and most likely the proposed new laws and rights will therefore be challenged by various actors in a society. Zwarteveen et al. (2005) propose four categories in which water laws and rights can be contested taking into account a plural legal perspective, namely:

1. The access to and control over water resources
2. The content and interpretation of water law determining the water distribution
3. The participation in decision making on water management
4. The discourses underlying the water law and implementation policies

This section outlines the four categories as interpreted by the authors. The first category refers to how physical access to, and control over, the finite water resource is negotiated and obtained in plural legal societies and on which basis. In negotiating access, water users actively utilise the various normative orders to legitimise their claims to water depending on which normative order serves their interests best, i.e. so-called forum shopping (Bentzon et al., 1998; Meinzen-Dick and Pradhan, 2005). However, not all normative orders have the same coercive means, nor do all enjoy the same degree of legitimacy (Von Benda- Beckmann, 1997; Spiertz, 2000; Von Benda-Beckmann, 2002). Actors with vested socio-political and economic powers can exert their stronger leverage position to influence which normative orders will prevail in the negotiations over water. Often the leverage position is closely linked to property ownership, such as infrastructure and land, as rights may become concretised rights over time, for instance, when rights become fixed in permanent concrete structures such as weirs, dams and field layouts (Mosse, 2008). However, powerful actors do not operate in isolation and the actual access to water resources is therefore an ongoing struggle between actors reflecting these social-political and economic interdependencies.

The second category is related to the conflicts and disagreements on the content of norms and laws and how they (should) determine rules and regulations on access to water. This category refers to the interpretation of water laws and the social meaning of water rights in society. It reflects the socio-economic power relations between the various actors and does not simply follow technical imperatives such as efficiency, but also reflects the historical and cultural values and ideas upheld in society. How water rights are understood and translated to rules for water use is coloured by the locally-accepted ways and traditions of dealing with water (Zwarteveen et al., 2005; Mosse, 2008).

The third category deals with the struggles over participation in decision making over water law and rights. Decision making spaces are often exclusive in the sense that some people are allowed to enter and participate in them and others not. Exclusion may be direct, based on class, gender or ethnicity. However, often exclusion is less direct and hidden in membership criteria, location of the meetings or language used. Moreover, being included in participation processes does not guarantee one's voice is heard as participation in decision making is determined by social relationships of power and dependency (Cleaver, 1999; Cornwall, 2003). Cultural norms associate certain forms of behaviour with knowledge and authority and others with ignorance, and in this way prescribe certain forms of behaviour to different social groups of people (Bentzon et al., 1998; Zwarteveen et al., 2005). This directly influences the self perceived capacity to participate in decision making and may even lead to self-exclusion (Wilson, 1999).

The last category in which water law and rights can be contested lies in the discourses used to articulate water problems and solutions. The way in which water problems and solutions are defined and conceptualised in a society is closely linked to the political agenda they promote (Molle, 2008; Mosse, 2008). Any understanding of water problems is based on representations and always implies a set of assumptions and (implicit) social and political choices. Knowledge produced on water is not merely neutral or scientific; it does not emerge by chance but, rather, is the emanation of complex webs of interests, ideologies and power as an inherent part of the water sector (Molle, 2008). Hence, the dynamics of water politics, including water law and rights, cannot be understood without also scrutinising the power relations, discourses and discursive practices that guide perceptions of water problems and proposed solutions. In an ever more globally-connected world order, this category includes analysing global political forces and global networks that influence the national policies on water (Conca, 2006; Mollinga, 2008; Swatuk, 2008).

3.3 Historical and institutional context of the catchment

The case study catchment is located in the Thukela River basin, in the foothills of the Drakenberg Mountains, located within the KwaZulu-Natal Province. The upstream part of the catchment is part of a former Zulu homeland and is inhabited by smallholder farmers. At the downstream end of the catchment commercial farms are located. This section describes the historical and institutional context of the catchment as well as the physical and socio-economic conditions in the case study area.

The Zulu tribe originates from Bantu communities and settled in the area in the 16th century. A crucial turning point in Zulu history occurred during the reign of Shaka (1816-1828). Prior to his rule, the Zulus consisted of numerous clans that were related but disorganised. During Shaka's reign conquered tribes were incorporated into the Zulu kingdom (Omer-Cooper, 1994). In 1653, the south-western part of South Africa was colonised by the Dutch and around 1825 the British settlers arrived on the east coast (Omer-Cooper, 1994; Wilson and Thompson, 1969). The Zulus fought several wars against the British, but surrendered in 1906. From then on the tribe was subjected, by European settlers and their descendents, to an increasingly harsh series of racist laws and practices that led to the disempowerment and subordination of the Zulus and other Black African tribes and which dispossessed them of access to natural resources (Mamdani, 1996).

Apartheid was a system of legalised racial segregation enforced by the White-dominated Government of South Africa between 1948 and 1994. Under apartheid, a series of measures were introduced as part of the policy of so-called 'separate development' that intended to create a South African society in which the White population would become the demographic majority. The creation of homelands was a central element of this strategy. Comprising no more than 14% of the country's area, ten arbitrary and often highly fragmented administrative territories were created in 1951 (Pickles and Weiner, 1991). These homelands were supposedly the original areas of settlement of what the state had identified as the country's main African ethnic groups and the Black population were made citizens of these homelands, denying them South African citizenship and voting rights (Mamdani, 1996). Within the homelands the Black Africans could aspire to self-rule under a chieftaincy (Ross, 1999). This subjected the inhabitants to the chiefs and made them lose access to ancestral land. Local tribal leaders were appointed by the Government to run the homelands, and uncooperative chiefs were forcibly deposed. By incorporating the traditional governance structure and paying salaries and other benefits to the traditional chiefs, the apartheid regime kept influence over the semi-autonomous homelands and could control resistance (Mamdani, 1996). Over time, a ruling Black elite emerged with a personal and financial interest in the preservation of the homelands. On advice of the apartheid Government large-scale reorganisation of the land use in the homelands was introduced under the Betterment Schemes. The reorganisation included dividing the land into distinct land-use zones, e.g., residential, arable and grazing areas. People living in the homelands were forced to move into demarcated residential zones and dispossessed of arable and grazing land. Only small plots were given to households to ensure the most basic crop production. The expressed goal of the Betterment Schemes was to 'rehabilitate' the land from the perils of overgrazing and 'inefficient' African land use, but, in reality, the Betterment Schemes facilitated increasing the population densities in the homelands (McCusker and Ramudzuli, 2007). The homelands became economically weak as the high population densities often far exceeded the carrying capacity of the land (Pickles and Weiner 1991; Ross, 1999). The education system was designed to prepare the Black population for manual labour and, with few local employment opportunities being available in the homelands, most men commuted to work on the commercial farms and in the mines of White South Africa (Bond, 2006). Women often stayed behind and were relegated to reproducing the future labour force and taking care of the sick and elderly (Penzhorn, 2005; Omer-Cooper, 1994). The apartheid politics sparked significant internal resistance. A series of uprisings and protests led to an armed resistance struggle against the White Government. Bloody armed clashes also occurred between opposing Black political parties, especially between the African National Congress (ANC) and the Zulu-dominated Inkatha Freedom Party (IFP). This violence, believed to be supported by the security forces of the apartheid Government, escalated at the end of apartheid. Today the tension between political parties still exists. Political apartheid was finally dismantled under internal and international pressure in a series of negotiations on the revision of the constitution from 1990 to 1993 (Omer-Cooper, 1994). The negotiation culminated in democratic general elections of 1994, which gave a landslide victory to the ANC.

Reconciliation of the society was the major concern of the new Government and it took on the transformation of the discriminatory legal systems. The Constitution was rewritten, as well as most laws, such as the National Water Act. As part of the institutional reform the Government structures were redefined and the homelands were dismantled, reincorporating their territory into the Republic. The national, provincial and local government levels all have legislative and executive authority in their own spheres and are defined in the South African Constitution as distinctive, interdependent and interrelated. The Constitution also acknowledges traditional governance structures and states that the country should be run on a system of cooperative governance (RSA, 1994). Prime advisory bodies of traditional leaders exist at all government levels and in the former homelands the traditional structures still play a formal executive role in addition to the local government structure (Lehman, 2007).

The case study catchment (Figure 3.1) occupies a total area of approximately 10 km^2 of hilly terrain with generally acidic soil. The mean annual rainfall is estimated to be 700 mm/yr and the estimated potential evaporation is between 1,600 and 2,000 mm/yr, at an elevation of about 1,250 m above sea level. A good drainage network has developed in the catchment with most of the streams being perennial. Extreme low flows occur in winter time between June and August (undisclosed reference).

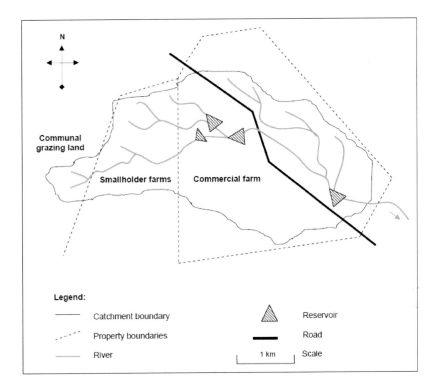

Figure 3.1: Sketch of the catchment.

The former Zulu homeland located in the upstream part of the catchment is mainly inhabited by smallholder farmers. The population in this part of the catchment fluctuates considerably as many commute to urban areas; however, it is estimated that around 500 people reside in the area on a permanent basis. Herding cattle and practicing agriculture are the main activities in the area. Box 3.2 provides details on the water sources used by the smallholder farmers. Cattle are kept for cultural reasons, although for a few farmers they also serve for commercial purposes. The cattle graze in summer time on communal land in the upper part of the catchment. Overgrazing has led to extensive soil erosion which has negatively influenced the natural water retention in the catchment (undisclosed reference). The agricultural plots are relatively small (0.5 to 2 ha) and the main crops grown are maize and beans for subsistence, although parts of the harvest are regularly sold. Supported by NGOs, an increasing number of smallholder farmers are growing vegetables in home gardens. The agricultural activities are not the main sources of income in the catchment as 37% of the households earn a regular income, 45% of the households have access to remittance from family members working elsewhere, and 82% of the households receive social grants from the Government (undisclosed reference), such as child support, old-age pension and disability grants (also available for HIV/Aids patients). The child support grant is ZAR 210 per child per month, while the latter two amount to ZAR 940 per month (SASSA, 2008). In addition, some smallholder farmers generate considerable income from illegally growing marijuana.

Box 3.2: Water sources used by smallholder farmers in the catchment

The smallholder farmers mainly use communal boreholes for domestic purposes. The boreholes are freely used and, even though few rules on water use exist, they are not adhered to. Not all boreholes work properly and some households are located more than 1 km from the boreholes. During the winter the farmers suffer from water shortages, sometimes even for domestic use when boreholes run dry. The cattle mainly drink from creeks and natural springs and some farmers have built small earthen dams on their plots to water their cattle. For agricultural activities the smallholder farmers primarily rely on rainfall. However, some home gardens are irrigated with water from the boreholes or springs. With support from NGOs, a few farmers have installed rainwater harvesting tanks, which collect runoff from the compounds and enable them to grow (supplementary) irrigated crops.

A commercial farm is located at the downstream end of the catchment with a total area of 1,560 ha. The commercial farm was established about 100 years ago and has been owned by three different families, all of British descent. The current farmer has owned the farm since 2002. The property includes four surface dams allowing the farmer to grow irrigated crops in both summer and winter seasons (see Figure 3.2). The farmer has registered his historical water use under the new water law, which means his water use is recognised as an existing lawful use (RSA, 1998a).

Figure 3.2: Photograph of the catchment, showing the houses and fields of the smallholder farmers in the forefront, and two reservoirs and irrigated lands of the commercial farm downstream (July, 2008)

Between the smallholder farmers and commercial farmer various direct links and interdependencies exist. The perennial streams in the former homeland replenish three of the four reservoirs of the commercial farmer downstream. However, the water carries considerable sediment loads caused by erosion in the upper part of the catchment. The sediments are trapped in the dams of the commercial farmer reducing their storage capacity.

During the winter months the cattle of the smallholder farmers roam the fallow fields and often trespass into the commercial farm. The cattle damage the crops of the commercial farmer and regular conflicts arise on this matter. In addition, interdependencies are created by the employment relationship: the commercial farmer employs approx. 30 permanent workers and up to 150 temporary workers during the harvest season. Most workers on the commercial farm are residents of the case study area while some others come from the surrounding areas.

The smallholder and commercial farmers residing in the catchment fall under the formal authority of the local municipality and are represented by an elected councillor at ward level in the municipal council. The composition of the municipal council is based on a mixed system of proportional representation and the constituency election system. The IFP is the ruling party in the municipality in which the catchment is located, while the ANC has the majority at provincial and national level. Amongst others, the municipality has the responsibility to ensure the provision of services (including water supply) to communities in a sustainable manner and to promote social and economic development (RSA, 1998b). In the former homeland, the traditional governance structure is still operational and the local chief controls access to land resources as custodian of the state-owned land. The land tenure reforms aiming at granting private ownership of land to the people living in the former homelands is highly contested by the traditional authorities (Lyne and Darroch, 2004). The traditional Zulu governance structure used to have committees of elderly men at village level and headmen at a higher spatial level as intermediaries under the chief. However, to make the traditional governance structure more compatible with the local municipality structure, the committees at village level in the case study catchment have been replaced in 2003 by elected leaders at ward level and elected councillors have been introduced in the council of the chief (RSA, 2003).

3.4 Contested water rights in the catchment

Although the National Water Act (1998a) directly aims at redressing the injustice of the past, inequities in access to water still exist in the catchment: so far the commercial farmer has kept his entitlements to the water and the water allocations to the smallholder farmers have not increased. This chapter presents the empirical data from the case study catchment and analyses the interactions between the various actors and the way in which they exert their water rights, following the framework presented in second section of this paper.

3.4.1   Category 1 – Access to and control over water

The current South African society is characterised by multiple legal realities. During apartheid, 2 different formal legal systems coexisted, the legal system in the White Republic and the legal system in the homelands. The legal system in the Republic focused on limiting power of the state and guaranteeing rights to its citizens, while the legal system in the homelands focused on strengthening the influence of the authorities to enforce customs upon its residents (Mamdani, 1996). After apartheid this was replaced by a new legal system; however, in contemporary South Africa the 2 legal systems of the apartheid era continue to exert their influence. In addition, as in every society, various normative orders, such as political ideologies, cultures and religions, coexist and influence in complex ways the context in which the negotiations over access to water take place. The Water Act is contested by various actors in society which are interlinked through social relationships and interdependencies. In the contemporary capitalist South African society the coercive means of the vested economic powers are strong (Bond, 2006; Swatuk, 2008). Large-scale water users such as commercial farmers, mining industries and electricity companies continue to receive water on economic grounds (Steyl et al., 2000). Research conducted in the northern parts of South Africa shows that wealthier people are better able to capture the available resources (Hope et al., 2004; see also Mosse, 2008; Sjaastad and Cousins, 2008). As a result, the highly stratified division of economic power along racial lines, and the status-quo of access to water resources in the study catchment is maintained.

The plural legal conditions in which access to resources has to be negotiated in former homelands is characterised by a more complex reality than in other parts of the country, as a result of the formal coexistence of the municipality and the traditional governance structure at local level. This coexistence determines that the municipality is responsible for water supply and the socio-economic development of the village (RSA, 1998b), while land tenure falls under the authority of the chief in study area. Close collaboration between the 2 institutions is required to ensure access to land with adequate opportunities to exploit water resources and sufficient tenure security to invest in hydraulic infrastructure. However, according to the interviewed farmers, in the case study area the institutions compete to increase their authority within the catchment. Only three of the eighteen interviewed smallholder farmers expressed trust in the local municipality whereas the other farmers regard it as being corrupt and dominated by struggles between the political parties. Almost all interviewed farmers are dissatisfied with the service delivery of the municipality. Ten interviewed smallholder farmers indicated that, as in the apartheid era, chiefs play an active role in politics. According to the smallholder farmers, chiefs are loyal to political parties in return for continuing support to legitimise their authority and other favours. In a young democracy with a large illiterate rural population voters can easily be manipulated and the chieftaincy fosters strongholds for the political parties. Four smallholder farmers indicated that the level of municipal service delivery to their community is low because the ruling chief belongs to the opposition party (ANC). They argue that, therefore, most smallholder farmers in the former homeland voted for the same opposition party; however, the elected councillor of the ward is from the ruling party (IFP).

The authority of the chief has lost legitimacy among the rural population as chiefs were seen as puppets of the White Government during apartheid (Mamdani, 1996). However, the current party politics and the state-introduced changes in traditional governance structure have also reduced the executive power of the traditional authority in the case study area. According to eight interviewed farmers, as a result of the imposed changes (e.g. the abandoning of the committee of elders) no institution has sufficient authority to spearhead joint initiatives at village level. For example, plans for jointly building small reservoirs for watering livestock and for irrigation have not materialised because of disputes over the location of the dams. Although access to land is controlled by the chief, the traditional structure does not have sufficient authority to make land available to the benefit of the community as a whole, and unused land is hardly available in the area as population density is high. The local politics leave the smallholder farmers in the catchment divided along party lines. The smallholder farmers indicated that village meetings are occasionally organised, but are mainly limited to conflict mitigation and providing top-down information, and are not used to discuss opportunities for progress or planning of joint initiatives. Although more than half of the smallholder farmers carry out joint activities, such as growing beans for sale and weaving mats, participation is limited to family members or political allies and these activities are often externally supported by NGOs.

Besides differences in political affiliations the composition of the community is also heterogeneous in terms of agricultural ambitions. Most smallholder farmers keep cattle for cultural purposes; however, only six farmers have the ambition to advance their crop-growing activities for commercial purposes. It should be noted that during apartheid considerable segments of the Black population had been de-urbanised during the forced relocation to the homelands (Mamdani, 1996). This could explain why agricultural ambition is relatively low in the former homeland. The heterogeneity of the community and associated difference in priorities gives the people who would like to farm a disadvantaged starting point in accessing municipal services which are often offered on a communal basis. It also negatively influences the negotiations over the redistribution of land and water with the downstream commercial farmer. Without a clear mandate to negotiate on behalf of the community the villagers are faced by a far more powerful opponent in the commercial farmers. The commercial farmer located at the downstream part of the catchment indicated that several smallholder farmers contacted him to discuss redistribution of natural resources: they requested access to part of his (fallow) land, receiving wood and support with constructing a dam upstream of his reservoirs. Instead, the commercial farmer has interest to jointly develop new water resources. However, according to him there is no established authority among the smallholder farmers with whom to thoroughly discuss the issue, and, in his opinion, the community is highly disorganised. In the absence of local government and/or the traditional authority to facilitate the negotiation process it leaves the smallholder farmers with virtually no opportunity to improve their access to the natural resources. Although, most likely, the legal plurality in the former homelands creates an enabling environment for certain groups to exert their claims, it also creates institutional chaos which prevents some smallholder farmers in the catchment from addressing the unequal access to natural resources and pursuing their agricultural ambitions.

3.4.2   Category 2 – Content and interpretation of water rights

The content of the National Water Act and how it should be interpreted is debated at various points. For example, the Act defines that water required to meet basic human needs is guaranteed as a right and all other water uses by humans are divided into priority categories (RSA, 1998a). The policy to implement this particular part of the Act identifies water for basic human needs only as water for domestic use and limits it to 25 litres per person per day. However, many smallholder farmers depend to a large extent on their own food production for their subsistence, and it can therefore be debated if and how water for subsistence farming should be included in the basic human needs (Hope et al., 2008). The guaranteed right of water used for environmental sustainability is also questioned (Swatuk, 2008) and is sometimes seen as a hobbyhorse of the White elite. Some smallholder farmers in the Olifants River Basin responded to water allocated to the environment with: 'As if they (read Whites) find fish more important than our lives' (Van Koppen and Jha, 2005: 209).

The Act also clearly indicates that the Government can reallocate water to redress the inequities in the society, but the reallocation is only legitimate if it is in the wider public interest (RSA, 1998a). However, it can be debated what the wider public interest is. Currently the Government recognises a need to 'balance' equity with productivity and profitability and is cautious about large-scale reallocations (Merrey, 2008). This approach is supported by the commercial farmer in the catchment, whose position is that he invests his own capital and takes financial risks to produce food to feed the rural population. He argues that if the land and water resources are given to the less experienced Black population, the agricultural production will significantly drop, which will negatively influence the national economy and is therefore not in the public interest. With the economic meltdown in neighbouring Zimbabwe, which, according to some politically-motivated voices is directly linked to the forced take-over of White-owned farms, most smallholder farmers in the catchment tend to agree with him; fourteen interviewed smallholder farmers acknowledge the role of the commercial farmers in the national economy and stress the need for a good relationship with the commercial farmers, regardless of the division of natural resources (see also Van der Zaag and Röling, 1996). The domination of fear in the debate through the comparison with Zimbabwe is a powerful tool in the hands of the commercial farmers to influence the societal meaning of 'the wider public interest', and in this way to maintain the status quo. However, three interviewed smallholder farmers explicitly indicated that a more fair distribution of natural resources is a prerequisite to improve the relationship with the commercial farmer downstream in the catchment. In addition, according to several smallholder farmers, the youth might not accept the current distribution of resources and threats to the commercial farmers are reported. An elderly female smallholder farmer (71 years) indicated that, in the absence of intimidating memories of the apartheid era and openly supported by some political movements, 'the youth might challenge the uneven relationship with the commercial farmers in the near future and might call for the Zimbabwean approach'.

3.4.3   Category 3 – Participation in decision making

The new legal system in South Africa makes a major shift in participation in decision-making processes in comparison to the apartheid era in which Black South Africans were excluded. Even though the formulation of the Water Act incorporated public views, and public participation during the implementation phase is stipulated, the decision-making spaces are still dominated by Whites and some Black elites (De Lange, 2004). Technical expertise and knowledge of water resource management is still White-dominated and the Government agencies have suffered greatly from 'brain drain' to private consultants as a result of the Black Economic Empowerment Act which focuses on achieving equity in (Government) employment (RSA, 2004). Nowadays the Government is highly dependent on hiring consultants to implement the Water Act and to facilitate public participation, while these consultants do not necessarily serve the interests of the Government nor are they necessarily familiar with the needs and desires of the Black community (Merrey, 2008). With 11 official languages in South Africa, language also plays a role in public participation, as most meeting are held in English and only partly translated in other languages (Van Koppen and Jha, 2005). High illiteracy rates in rural South Africa seriously hamper the involvement of rural communities (Simpungwe, 2006).

For public participation to be effective in decision making, liaison with interest groups is essential. Within the participation process the rural communities are often categorised as one single interest group, for instance, for their representation in the water user associations that will be established under the Water Act (authors' unpublished data). The smallholder farmers in the catchment are faced with the challenge that they are often expected to act on a communal basis in participation processes while they are strongly divided amongst themselves. Underlying causes of the difficulties for the community to act collectively are the weak family ties as a result of the resettlement policies and the condemnation of every form of self-organisation during apartheid. Moreover, the perception of dependency on external support (e.g. Government and NGOs) for development is strong, as fourteen smallholder farmers indicated that external institutions have to take care of their (basic) needs. The (limited) monetary resources available in the community are barely invested in productive activities and confidence in their own entrepreneurial skills is low. When asked what they do for a living more than half of the smallholder farmers indicated that they do not have an occupation (any longer) and rely on social grants from the Government. Only four out of eighteen smallholder farmers acknowledge their own agricultural activities, even among the smallholder farmers who regularly sell their surplus. Potentially the dependency created during apartheid undermines the self-reliance of the smallholder farmers and limits their self-esteem to participate in decision making on water resources.

The scars from the past still dominate the relationship with the downstream commercial farmer. Although various interdependencies exist between the smallholder farmers and the commercial farmer and both parties would benefit from a good relationship, the smallholder farmers feel inferior and believe they depend more on him than the other way around. Although the commercial farmer realises that he is the minority in the area and his farm could be occupied overnight, he also believes the smallholder farmers will not take the risk to illegally challenge the current situation as he employs a considerable number of community members on his farm. For water, the commercial farmer is dependent on the smallholder farmers as most of the water in his reservoirs originate from their area; nevertheless he indicated that his current water entitlement gives him sufficient confidence that he will continue to receive water in the future. For reallocation of water under the Water Act downstream lawful uses have to be considered and compensated if negatively affected, unless it can be proven that the reallocation is to *"rectify an unfair or disproportionate water use"* (RSA, 1998a: 22-7). Nevertheless the commercial farmer trusts that he has a strong case to oppose potential redistribution of water in the catchment, which worries the smallholder farmers who are less familiar with the content of the Water Act. The understanding of each other's realities and respect for cultural values is limited. The smallholder farmers indicated that they have invited the commercial farmer, in vain, to several official ceremonies in the community to strengthen the bond between them. The commercial farmer indicated that he does not have time to attend the time-consuming ceremonies and prefers to limit their interaction to purely business matters. The local government is absent in bringing the parties together to facilitate the reconciliation process and foster collaboration. Currently, communication between the parties is restricted to conflict mitigation over cattle trespassing. Five smallholder farmers indicated that, out of frustration over the current situation, people living in the case study area sabotage the activities of the commercial farmer, e.g., deliberate trespassing of cattle on the commercial farm, stealing of crops, and destroying equipment. Although it affects his business, the commercial farmer indicated that he does not act on this impairment but tolerates it to avoid it becoming worse. Both parties pointed out that potential collaboration on the development of alternative water resources is seriously affected by the current negative relationship.

The Water Act is explicit in its aim to redress the inequities based on gender (RSA, 1998a) and mainly refers to providing domestic and productive water to poor women in the rural areas. However, some cultural values obstruct participation of women in decision making. The Zulu culture is patrilineal in which the customary heritage of property and power positions is through the male lineage, from father to son (Mair, 1969). In the Zulu culture women have limited political status, and resources such as land and livestock are mainly owned by men. Although water resources are mainly used by women for domestic purposes, and growing crops is regarded a female activity, women are not involved in the maintenance and future planning of the water resources and have a low political status (Penzhorn, 2005). Three female smallholder farmers indicated that they would like to be involved in decision making at village level, but, for community meetings, often only men are invited. The few women who do attend the meetings on their own initiative indicated that they have difficulties in speaking up and being heard. In particular the presence of male relatives of their husbands restricts them from raising their voices. Half of the interviewed women in the catchment are not aware of the institutional structures in their community and indicated that men do not inform them about governance issues. Two male farmers responded that the village meetings focus primarily on issues related to cattle which they regard as a male business. Joint initiatives of women to increase their influence in the community do not take place, according to the interviewed female farmers, because the women in the community are not united. Some NGO-supported agricultural groups include women (undisclosed reference); nevertheless, the decision-making spaces remain strongly male-dominated.

3.4.4 Category 4 – Discourses underlying water law and implementation policies

The National Water Act (RSA, 1998a) was politically driven by the need to redress the inequities in society and to create equal opportunities for all citizens. However, this socialist-oriented political ideology of the new Government was not compatible with the capitalist principles on which the South African economy was based (Bond, 2005; Hart, 2006). Maintaining a strong economy through focus on market efficiency, competition and productivity was supported by the economic elite in South Africa as well as by the geopolitical agenda (Hart, 2008). Through conditional policy reforms attached to loans from the World Bank and IMF the new Government was persuaded to adopt a neo-liberal approach (Bond, 2005; Harvey, 2005). The neoliberal doctrine suggests that human well-being can best be advanced by liberating individual entrepreneurial freedoms and skills in an institutional framework characterised by free markets and globalisation of trade (Harvey, 2005). It proposes limiting the control of the Government on the economy, as well as the privatisation of public services and property rights over natural resources (Harvey, 2005; Ahlers and Zwarteveen, 2009). Based on these neoliberal principles the South African Government introduced the Growth, Employment and Redistribution (GEAR) policy in 1996, which focused on achieving equity by enlarging the 'wealth' cake. However, in a country characterised by structural inequalities along racial lines in terms of educational background and access to resources, emphasis on market competition between actors is unlikely to lead to equity. Moreover, shifting power from the state to the market and focusing on productivity seriously hampers the redistribution of natural resources to the previously-dispossessed groups in the society. According to Bond (2005), neo-liberal policies have amplified rather than corrected the economic distortions created during apartheid and it is argued that the neo-liberal doctrine has replaced racial segregation under apartheid with class segregation in contemporary South Africa (Bond, 2006; Harvey, 2005; Swatuk, 2008).

The dominance of the neoliberal discourse becomes visible in the chosen approach for the implementation of the National Water Act and the priorities set for the water allocations (Bond, 2006). For example, the choice of a sectoral approach to water delivery might be suitable for the high-volume users in the better-served areas, but in the rural areas water resources are often used for multiple purposes and integrated service delivery would be more effective (Van Koppen and Jha, 2005). The smallholder farmers in the catchment indicated that they have to deal with different Government departments for their water supply depending on the purpose of the water use (e.g. domestic, various productive uses), while in practice the same water resources are being used. Increasing access to the water resource is therefore challenging, as it needs collaboration between the various Government departments. Moreover, it remains unclear in the Water Act under which category water use for small-scale commercial production falls (Perret, 2001). Water abstraction for subsistence farming is permitted under the Water Act without registration or payment (RSA, 1998a), while for all other water uses a licence needs to be granted. However, the smallholder farmers in the case study catchment often only sell part of their harvest depending on the yields and market opportunities, which makes the water use for commercial purposes difficult to predict, and is, hence, licensed. Furthermore, to obtain a licence the Water Act stipulates that the water use should be in line with the catchment management strategy negotiated within the new water user platforms (RSA, 1998a). In the absence of organisational structures around water in the former homelands, meaningful representation in these new arenas will be problematic, which leaves the emerging individual smallholder farmers formally without rights to water for commercial uses and favours the established well-organised large-scale commercial farmers in the catchment. This is now acknowledged by the South African Government and the proposed granting of a general authorisation for small-scale commercial water use is currently debated. The water allocations at national level also clearly demonstrate the underlying neoliberal discourse: even though the smallholder farmers in the case study area indicated that a shortage of water limits their economic development, the Thukela River Basin is transferring approximately 75% of its surface water yield to adjacent river basins to support commercial activities of industries and hydro-power production in urban parts of the country (DWAF, 2003). Consultants hired by the Government to study the water availability and water use in the Thukela River Basin recommended continuation of water transfers to other parts of the country as, in their view, no strong economic drivers within the basin exist to stimulate development. They regard water resource development for the sole purpose of irrigation economically unviable and recommend allocating the water to other sectors (DWAF, 2003).

Even more striking is the lack of an integrated approach across sectors, which disconnects the water reform process from the land reform process. In South Africa the inequity in land and water resources is closely linked, with the smallholder farmers mainly relying on green water resources (Rockström et al., 1999; Savenije, 1999). Significant increases in access to water can therefore not be achieved in the overpopulated former homelands unless the inequity in land distribution is addressed simultaneously (Hope et al., 2004; 2008). Currently the distribution of land under the land reform program,e is based on the neoliberal 'willing seller, willing buyer' principle, in which land is bought in conformity to market prices (DALA, 2006). However, this approach results in a slow pace of the land reform process (Lyne and Darroch, 2004; Lahiff and Cousins, 2005; Hart, 2006; Cousins, 2007; Peters, 2009) and, according to the smallholder farmers in the case study catchment, only fragmented pieces of communal land become available with often limited access to (blue) water resources. For instance, through the land distribution programme the smallholder farmers have obtained approximately 600 ha of dry mountainous land adjacent to their community, and another piece of land that was offered to them was more than 10 km away from their current location. Moreover, it is slowly being acknowledged that, based on market prices, the Government will never be able to afford to buy sufficient land to radically address the existing inequity (DALA, 2006).

Finally, in the South African Constitution it is stated that the country should be run on a system of cooperative governance between the Government and the traditional governance structures (RSA, 1994). This principle is admirable from the reconciliation perspective; however, besides the additional challenges it creates in the former homelands, as described above, it is to a certain extent also a facade. With the Government having the constitutional authority, the democracy discourse is dominant and the role of the traditional authorities is limited to an advisory role, despite the protests of the traditional leaders (Lehman, 2007). In this way the cooperative governance system reproduces the inherited inequalities between the

Government and the traditional authorities. The dominant democratic discourse becomes further apparent in the manipulation of the traditional structure to comply with Government structures and the requirements for democratically-elected representation in the traditional governance structure (RSA, 2003), which has affected the executive power of the traditional authorities. Nonetheless, it can be debated as to who the traditional governance structure is representing, particularly after the manipulations by the apartheid regime (Mamdani, 1996). In the traditional governance structure the Black elite might be represented, but it can be questioned whether the rural poor are represented (Sjaastad and Cousins, 2008). It can therefore be argued that the adoption of a cooperative governance structure was a politically-motivated choice to support the Black elite rather than a prerequisite for the socio-economic development of the rural areas.

3.5 Discussion and conclusions

Although the National Water Act (RSA, 1998a) is comprehensive in its legislation and provides powerful legal tools to address poverty eradication and redress inequities inherited from the past, in reality little transition in the access to and control over water resources has been achieved (Bond, 2006; Cullis and Van Koppen, 2009; Merrey, 2008). This paper emphasises that water law on paper is not sufficient, as it is not implemented and enforced in a vacuum, but in a society thick with historically entrenched socio-economic and political inequities. The transition from apartheid to post-apartheid South Africa has been characterised by a negotiated transformation with emphasis on reconciliation. As a consequence the legacy of apartheid and the nature of the transitional arrangement still determine, to a large extent, today's political and economical reality, with the elite, White and some Black, in charge. Neo-liberal geopolitics has left the South African Government faced with the challenge to redress access to natural resources in an era characterised by the promotion of the private sector over the public sector (see also Lahiff, 2003; Hart, 2006; Swatuk, 2008). Hence, the Water Act is highly contested by the vested economic elite and up to now the status quo of an unfair distribution of water along racial lines is maintained, leaving the smallholder farmers in the study catchment managing in the margin.

The Water Act provides opportunities for citizens to contest unequal access to water resources through a bottom-up approach. However, smallholder farmers in the case study catchment face various challenges that prevent them from claiming their rights. These can be summarised in four points. First, the institutional chaos created in the former homeland as a result of the formal plural governance structure directly influences the executive power and legitimacy of the various authorities, thwarts collaboration, and creates 'fuzziness'. Second, the tendency of authorities and NGOs to approach the smallholder farmers as a united community and offer services and resources on communal basis is problematic, as the farmers are divided and heterogeneous in their ambitions, which makes it difficult for individual farmers to advance their agricultural activities. Besides, it can be questioned if the residents in the former homeland should be approached as farmers, as a substantial part of the residents do not perceive themselves as such. This is in line with the statement of Dlali (2008: 44) 'most people currently living in the rural areas in South Africa are 'rural dwellers' rather than farmers'. Third, the disconnect between the water and land reform programs puts the smallholder farmers in the dilemma of acquiring land without (blue) water or water without land. Fourth, the low self-esteem as a result of the structural, racial and gender inequalities hinders the smallholder farmers from claiming their rights and challenging the unequal access to water for (productive) uses. The smallholder farmers adopt an underdog position and their attitude tends to be submissive, especially the female smallholder farmers. As argued by

Zwarteveen et al. (2005), historically entrenched markers of behaviour that have served for generations to delineate and express the differentiation based on race and gender are not easily undone through legal changes.

The analytical framework used in this paper distinguishes four categories in which water law can be contested, based on Zwarteveen et al. (2005). The framework has been useful to comprehend the challenges met in the reform process, as it systematically analyses how water law is contested even if no explicit conflicts over or open claims on water exist. It reveals the different manifestations of struggles over access to water resources and how they are linked to self-perceived capacity of the actors to challenge or maintain the existing situation. In this way, the framework exposes how the implementation of the new water law is dynamically linked to deeper entrenched power structures in society and details how actors actively use and/or 'misuse' the water law to shape the negotiations over water. However, the framework focuses on water law, which only deals with water allocation and not with water distribution. Therefore, to increase the physical access of the smallholder farmers to (productive) water, distributional issues, such as the availability of hydraulic infrastructure and the capacity to manage, maintain and operate the infrastructure, need to be tackled simultaneously (Van der Zaag and Bolding, 2009). Investment in hydraulic infrastructure is closely linked to land availability and land tenure. Hence, this emphasises the importance of integrated and cross-sectoral approaches, as well as the importance of reviewing the cooperative governance structure in the former homelands. However, as argued before, water systems are not only shaped by, but also themselves shape and reinforce, social-political and economic relations, and historical analysis shows that hydraulic infrastructure has been used by wealthier farmers to assert their claims by ensuring that their 'rights' are fixed in permanent concrete structures (Mosse, 2008). Challenges associated with development of hydraulic infrastructure should therefore not withhold the Government from investing in hydraulic infrastructure in the previous disadvantaged former homelands.

In conclusion, the Water Act and other post-apartheid laws raised expectations amongst the various groups in society and gave hope for a more prosperous future, but after a decade of reconciliation the division of wealth is hardly unchanged (Swatuk, 2008). The ongoing protests and eruptions of violence in South Africa are resurgences of the underlying struggle over the structural inequities in society; the smallholder farmers in the case study area indicated that youth from their community had been involved in mid-2008 in xenophobic attacks (Mail & Guardian, 2008), and they believe that the violence was an expression of the frustration about the dire conditions in which people live compared to the living conditions of the elite (see also Hart, 2006; Swatuk, 2008; Peters, 2009). The difficulties encountered in the water reform process are illustrative of what is happening in the society at large. Although so far the elite have been able to maintain the status quo in the division of wealth, it can be questioned how robust the situation is, with the (potentially explosive) undercurrents in the society, and, hence, what price is being paid to maintain this status-quo?

4. **The question of inclusion and representation in rural South Africa: challenging the concept of water user associations as a vehicle for transformation** [39]

Abstract

The promotion of local governance and the transfer of water management responsibilities to water user associations (WUAs) have been central in water reform processes throughout the world, including in the reforms that took place in post-apartheid South Africa. This paper reflects on the notions of inclusion and representation as put forward by the various actors involved in the establishment of a WUA in a tertiary catchment in the Thukela River Basin. The paper describes how the WUA in the study catchment came to be dominated by commercial farmers, despite the South African government's aim to redress the inequities of the past by the inclusion and representation of historically disadvantaged individuals. The authors argue that the notions of inclusion and representation as embedded in the concept of the WUA are highly contested and more aligned with the institutional settings familiar to the commercial farmers. The paper concludes that, unless the inherently political nature of the participatory process is recognized and the different institutional settings become part of the negotiation process of the 'why' and the 'how' of progressive collaboration at catchment level, the establishment of the WUA in the study catchment will not contribute to achieving the envisioned transformation.

---

[39] This chapter is based on: Kemerink, J.S., L.E. Méndez Barrientos, R. Ahlers, P. Wester, P. van der Zaag (2013) Challenging the concept of Water User Associations as the vehicle for transformation: the question of inclusion and representation in rural South Africa. *Water Policy* 15(2): 243-257

4.1 Introduction

The promotion of local governance and the transfer of water management responsibilities to user groups, commonly referred to as water user associations (WUAs), has been central to water reform processes throughout the world since a more institutional and integrated approach to water management was introduced in the 1980s (Cleaver, 2002; Meinzen-Dick and Pradhan, 2002; Molle, 2004). Especially within the irrigation sector, the involvement of water users in the management of irrigation systems has become common practice (Uphoff et al., 1999; Mollinga and Bolding, 2004; Rap, 2006; Merrey et al., 2007). The participation of user groups is considered to be the way to operationalize decentralization for democratic transformation and to achieve empowerment (Cornwall, 2003, 2007). As Cleaver argues *"participation has become an act of faith in development; something we believe in and rarely question"* (Cleaver, 1999: 597). Embracing this paradigm, the post-apartheid South African government defined participation as one of its cornerstones to redress the racist water policies of the past. The National Water Act (1998) recognizes that 'while water is a natural resource that belongs to all people, the discriminatory laws and practices of the past have prevented equal access to water and use of water resources' (RSA, 1998: second preamble). Within this specific problem framing, the purpose of the Act is defined as follows:

> *"to ensure that the nation's water resources are protected, used, developed, conserved, managed and controlled in ways which take into account amongst other factors: (a) meeting the basic human needs of present and future generations; (b) promoting equitable access to water; (c) redressing the results of past racial and gender discrimination; ..."* (RSA, 1998: 2)

The Act continues by expressing the need for participation *"and for achieving this purpose, to establish suitable institutions and to ensure that they have appropriate community, racial and gender representation"* (RSA, 1998: 2). However, despite the significant claims, there is little empirical evidence of the long-term effectiveness of participation in materially improving the conditions of the most vulnerable people or as a strategy for achieving social change by giving voice to the previously excluded (Mayoux, 1995; Cleaver, 1999; Cornwall, 2003; Williams, 2004; Goldin, 2010). So what is it that makes WUAs the appropriate vehicles to contribute to the transformation of the water sector as envisioned by the National Water Act?

This paper aims to examine this question and thereby contribute to the ongoing discussion of the water reform process in South Africa (van Koppen and Jha, 2005; Waalewijn et al., 2005; Merrey et al., 2009; Goldin, 2010; Brown, 2011; Kemerink et al., 2011; Movik, 2011; Schreiner and Hassan, 2011; Bourblanc, 2012). The paper presents empirical data on the establishment of a WUA and analyses the impact on the access to and control over water resources for the various groups involved. The catchment used as a case study for this paper is located in the Thukela River Basin in the south eastern part of South Africa. The actual name and location of the catchment will not be revealed due to ongoing political sensitivities between various actors in the case study area. The findings presented are based on in- depth semi-structured interviews with 38 residents of the former homelands and 18 commercial farmers within the catchment carried out between June 2008 and July 2011. The interviews addressed amongst other issues such things as personal histories, livelihoods, access to and use of water and land resources, involvement in water management organizations, and perceptions of the water and land reforms. The interviewees were selected by a stratified random selection procedure to guarantee geographical spread and to obtain input from various age, race, class and gender groups. The findings of the interviews were cross checked through focus group discussions, observations, comparison with existing literature and by consultations with resource persons such as representatives of local authorities, government officials and non-governmental organizations (NGOs) active in the region.

This paper first explores the theoretical considerations used in the participation paradigm. In the next section the catchment is described, including an analysis of the historical and institutional context. The process for the establishment of the WUA as set down on paper and as implemented in practice in the case-study catchment is then narrated. This is followed by a critical analysis of the participation process in terms of inclusion and representation. In the concluding section, the general concept of WUAs is discussed in terms of its (potential) role in achieving transformation of the South African water sector.

4.2 Theoretical considerations

As pointed out by Cleaver (1999), participation has become a paradigm in managing water resources and WUAs are seen as the platforms for structuring stakeholder involvement. Government and development agencies in charge of establishing WUAs tend to focus on getting the techniques for the participatory process right, while at the same time trying to conceal the political issues at stake. As long as 'all' water users are 'included' through some form of 'representation' it is assumed that participation will automatically lead to 'better' water management practices, or even to the improved material conditions of the most vulnerable groups in society (see also Wester et al., 2003). But what do inclusion and representation mean, and who ought to define these notions? These are inherently political questions and empirical evidence shows that ignoring this contested nature of the participation process may imply that structural change (in terms of equity) will not automatically be achieved (Mayoux, 1995; Cleaver, 1999; Manor, 2004; Ahlers and Zwarteveen, 2009; Goldin, 2010).

As an analytical framework, this paper adopts the view that platforms on which the negotiations over water take place are characterized by plural legal conditions. Legal pluralism refers to the coexistence and interaction between different normative orders in the same socio-political space (Von Benda- Beckmann, 1997; Boelens et al., 2005; Von-Benda-Beckmann and Von-Benda-Beckmann, 2006). A normative order can be defined as any system of rules or shared expectations of what people should or should not think, say or do concerning a particular situation. In legal pluralism the different normative orders may originate from various sources such as political ideologies, economic dogmas, knowledge regimes, religions and cultures at different spatial and temporal scales. Even though legal pluralism seems an abstract concept, in daily life we all deal with legal plural conditions: for instance, we prepare our food according to our religious beliefs, we interact with our families based on cultural traditions and we provide labour conforming to the current economic doctrine. The different normative orders in society can be complementary, overlapping or even contradictory, creating space for bargaining and manipulation by different water users. This generates a plethora of (hybrid) local rules and arrangements (Meinzen- Dick and Pradhan, 2005) which in their very essence are shaped by history and embedded in local realities.

Viewing the introduction of new platforms for interaction, such as WUAs, from a legal plural perspective involves opening up space for bargaining not only over water use per se but also over the institutions that govern these interactions. How to define inclusion and representation becomes an integral part of the negotiation process: who will be included, on what, when and where? Who will represent who, on which basis and how? Moreover, the new platforms will not automatically replace the existing domains of interaction: empirical evidence shows that water resource management can take place almost entirely outside the WUA structures through practices embedded in social networks, daily interactions and the application of cultural norms (Cleaver, 1999). This coexistence of different domains raises questions about which domains are visible and which remain invisible, as well as which domains are formally recognized and which are not. As the state formally recognizes the WUAs as the main decision-making platforms for water at a local level, they have become important negotiation domains since they constitute bodies that may legitimize contested claims to water. However, the interaction between old and new domains cannot be ignored, as authority and the bureaucratic apparatus required only develop over time. Hence, notions of inclusion and representation need to be analysed in the context of various interlinked and politicized domains of interaction that deal with water and related issues (see also Warner et al., 2008).

With these theoretical considerations in mind, this paper seeks to understand the impact of the establishment of WUAs on collaborative efforts at catchment level by critically analysing the notions of inclusion and representation as put forward by the different stakeholders involved. On this basis, the paper discusses the potential of the WUA concept as a potential vehicle to contribute to the transformation of the South African water sector.

4.3 Setting the scene

The case study area is a tertiary catchment of about 1,500 km^2 within the Thukela River Basin located in KwaZulu-Natal province, in the south eastern part of South Africa. The catchment is located in the foothills of the Drakensberg Mountains and has three main tributaries that flow eastward from a steep escarpment across low mountains to the lowlands where they join and flow into the Thukela river. The rainfall varies considerably from up to 1,000 mm/yr in the upstream mountainous areas to 640 mm/yr in the lowlands. The estimated potential evaporation is between 1,600 and 2,000 mm/yr at an elevation of about 1,250 m above sea level. Most of the streams are perennial with extreme low flows in winter (between June and August).

The catchment is primarily inhabited by two distinct groups: commercial farmers of European descent residing in the lower parts of the catchment and communities from the Zulu tribe in the upstream parts of the catchment. The segregation between these two groups is a direct result of the discriminatory policies introduced by the British settlers' colonial authority (1906–1948) and further elaborated and imposed by the Afrikaner-led apartheid government (1948–1994) (Omer-Cooper, 1994; Mamdani, 1996). Under these policies, the black African ethnic groups, including the Zulu tribe, were dispossessed from their access to natural resources and relocated to so-called homelands (Pickles and Weiner, 1991). In the homelands, residents became subjected to chiefs, appointed and paid by the apartheid government, who had personal and financial interests in the preservation of the homelands (Mamdani, 1996). Within the homelands, residents were forced to move into demarcated residential zones and only small plots (0.5–2 hectares) were given to households to ensure basic crop production based on rainfed agriculture. This strict spatial planning facilitated the process of relocating more people into the homelands (McCusker and Ramudzuli, 2007), with the result that the homelands became economically weak as population densities exceeded the carrying capacity of the land (Pickles and Weiner, 1991; Ross, 1999). In this way, the homelands became cheap labour pools for white businesses, with most men commuting to work for the white-owned commercial farms and mines, while women stayed behind nursing children and cultivating the small plots for subsistence (Omer-Cooper, 1994; Penzhorn, 2005).

In the meantime, the white minority enjoyed support from the government to acquire large parts of land and to construct hydraulic infrastructure, such as dams and weirs, in order to establish sophisticated irrigation systems to support commercial agriculture (Chikozho, 2008). In the catchment, the irrigation of wheat, soya beans and maize by commercial farmers formed the major water use throughout the year, though irrigated fodder crops and pastures for livestock also claimed part of the available water resources. In response to increased competition over water, commercial farmers started to organize themselves around water at the start of the 20th century and irrigation boards were initiated under the Water Act of 1926. Four irrigation boards were formed within the study catchment: one along each tributary and a fourth one downstream along an irrigation channel. Over the years, the irrigation boards' command area increased so that today the four irrigation boards together manage a total of 6,500 hectares of irrigated land belonging to 84 farmers. The irrigation boards own the dams that have collectively been built by their members, though the rights to the stored water depend on the farmers' individual contribution to the construction of the infrastructure. This has created a complex and innovative system of water sharing arrangements supported by refined institutional structures within the irrigation boards (Méndez, 2010; Méndez et al., forthcoming)[40].

Political apartheid was dismantled after internal and international pressure in a series of negotiations over the revision of the constitution from 1990 to 1993 (Omer-Cooper, 1994). This culminated in the democratic general elections of 1994, after which the new government took on the transformation of the discriminatory legal systems as its prime objective. As part of the institutional reform, the government structures were redefined and the homelands were dismantled, reincorporating their territory into the republic, and comprehensive land reforms have been initiated (Cousins, 2007). Almost two decades later, the legacy of apartheid is still clearly visible in the study catchment with its large white-owned commercial farms, relatively crowded former homelands and impoverished urban townships. The current land holdings of the commercial farmers in the catchment range between 30 and 1,500 hectares with private dams and sophisticated hydraulic infrastructure for irrigation as well as pastures for livestock (Méndez, 2010). The residents of the former homelands have access to plots with sizes between 0.5 and 4 hectares that mainly depend on rainfall with which to cultivate maize and beans (Méndez, 2010; Kemerink et al., 2011). The limited changes in access to land and water that have taken place since the start of the reform processes in the study catchment are summarized in Table 4.1.

---

[40] Méndez et al. (*forthcoming*) is included in Annex 2.

Table 4.1: Acquired land resources and water entitlements under reform processes in the study catchment.

	Acquired land resources	Acquired water entitlements	Beneficiaries
Commercial farmers	None	Dam storing 4 million m³ of water	7 families
		Dam storing 3 million m³ of water	41 families
Residents of former homelands	500 hectares residential area + 400 hectares farming land	Water permit to irrigate 100 hectares	200 families
	600 hectares grazing land	None	77 families

It is within this context that the water reform process is taking place through the implementation of the National Water Act (RSA, 1998). The Act is widely recognized in policy circles as one of the most comprehensive and progressive water laws in the world (Biggs et al., 2008; Merrey et al., 2009). It defines the state as the custodian of the nation's water resources and only water required to meet basic human needs and to maintain environmental sustainability is guaranteed as a right (RSA, 1998). This fundamentally moves away from the previous water acts which were largely based on riparian water rights. Moreover, the new Water Act gives the state a strong tool to redress race and gender inequities inherited from the past (Van Koppen and Jha, 2005). The Act calls for extensive institutional reforms within the water sector based on the principle of decentralization, with the establishment of WUAs as the prime bodies to facilitate stakeholder participation at a local level.

4.4 Establishment of water user associations

4.4.1 Process on paper

A central part of the National Water Act was the establishment of Catchment Management Agencies (CMAs) to develop strategies for the use and protection of the water resources in each of the nineteen identified water management areas in the country. These strategies should include a water allocation plan that defines the principles for allocating water to existing and prospective users, taking into account all matters relevant to the protection, use, development, conservation, management and control of water resources (RSA, 1998). Under the CMA, WUAs are to be established at local level, with the primary role of undertaking water-related activities for the mutual benefit of their members, including supervision and regulation of water distribution and construction, and operation of hydraulic infrastructure. The Water Act indicates that existing water boards, such as the irrigation boards of commercial farmers, are primary points of departure for the establishment of the WUAs: *"An (irrigation) board continues to exist until it is declared to be a WUA in terms of subsection (6) or until it is disestablished in terms of the law by or under which it was established"* (RSA, 1998: 2).

The policies to implement the decentralization of water resources management indicated that the CMAs would be established in 1999 and the WUAs would follow in 2000 (RSA, 1998).

To facilitate the establishment of the WUAs, an 'Irrigation Board Transformation Guideline' was written by the Department of Water Affairs (DWA). The guideline emphasized the need for appropriate representation of historically disadvantaged individuals[41] in terms of race and gender in the management committees of the WUAs. The guideline specified that:

"The transformation process also requires that other imbalances within the area of operation of a WUA be addressed. The process should, amongst other things, aim to: (1) avoid a situation where one group is being dominated by another; (2) ensure representation for minority groups; and (3) assist in resolving conflict by creating balanced representation in terms of the various categories of users" (DWA, 2000: 18)

Aware of the fact that most historically disadvantaged individuals are currently not relevant water users for agricultural purposes because of their limited access to land, water and infrastructure, the DWA made explicit that *"domestic water users will in most cases be an interest group of sufficient significance to justify a nominated representative on the management committee"* (DWA, 2000: 17). The guideline also indicates the possibility of enlarging the area under control by the irrigation boards in order to include upstream and downstream communities of historically disadvantaged users (RSA, 1998).

4.4.2 Process in practice

DWA has encountered enormous delays in operationalizing the decentralization of water resource management. So far, eight CMAs have been established of which only two can be considered to be functional, both with limited success (van Koppen and Jha, 2005; Waalewijn et al., 2005; Merrey et al., 2009; Karar et al., 2011; Bourblanc, 2012); the Thukela Basin continues to be managed under the old centralized system. Nevertheless, DWA decided to go ahead with the establishment of the WUAs by sending the Irrigation Board Transformation Guideline to the four irrigation boards in the case study area in 2000, which marked the start of the official process.

According to the Water Act, each irrigation board would be restructured into a WUA. However, the four irrigation boards in the case study area proposed to DWA that they form one WUA instead of four separate WUAs. According to the commercial farmers, this was suggested because the four irrigation boards are located along interconnected tributaries in one tertiary catchment, so together they represent an integrated hydrological unit. Moreover, they share hydraulic infrastructure and associated complex administrative and financial systems (Méndez, 2010; Méndez et al., *forthcoming*). Eventually, after several discussions and negotiations with the commercial farmers, DWA allowed the establishment of a single WUA for the catchment.

The Water Act stipulates that 'any person holding office with a (irrigation) board when this Act commences continues in office for the term of that person's appointment' (RSA, 1998: 3c), which meant that the four chairmen of the irrigation boards became de facto members of the management committee of the WUA. These founding members drafted the constitution of the WUA based on example constitutions: *"We got the constitution from another WUA I*

---

[41] An 'historically disadvantaged individual' is a policy term in the South African context that refers to any person, category of persons or community who is disadvantaged by unfair discrimination before the Constitution of the Republic of South Africa prior to 1993 (Act 200 of 1993) including women and individuals with a disability (DA, 2004).

guess, I never read it, it is a thick document and the WUA Secretary just replaced their names with ours" (Interview CF1, 2011). The constitution details the structure of the WUA, its governing laws, as well as the three objectives of the WUA, which are:

1. to manage and promote efficient, equitable and sustainable use and distribution of water resources and water works;
2. to strive to ensure appropriate community, racial and gender representation and participation in the affairs of the Association; and
3. to control water development within the area of operation.

After drafting the constitution a public awareness campaign was organized jointly by the four irrigation boards, with the placing of advertisements in local newspapers to announce the new WUA and invite community members residing in the municipal area to join the transformation process. On request of DWA, the advertisements were republished to increase the publicity and, in 2009, a first meeting was held with the assistance of DWA officials. About sixty historically disadvantaged individuals[42] attended the meeting, which was held in English with summarized translations in Zulu. The subsequent meetings were without DWA's involvement and the attendance of historically disadvantaged individuals dramatically reduced: at the second meeting only fifteen of them were present and, by the end of 2009, the number was reduced to eight. Out of the eight people, four did not reside within the hydrological boundaries of the WUA (though within the advertised municipal area). Finally, two of the four remaining eligible historically disadvantaged individuals were elected, together with the re-election of the other members, to become members of the management committee of the WUA by the people attending the meeting. In this way one woman from an urban settlement took up the role of gender representative and one member of a community that had recently acquired a commercial farm became the representative of the emerging farmers[43]. DWA accepted the outcomes of the elections without further scrutinizing the process and arranged multiple training sessions on gender for the WUA management committee. These training sessions aimed to establish a common understanding of how socially-constructed gender relations affect water management practices. Why racial relations (which noticeably influence water management practices in the catchment) were left out remains unclear. While emerging sector representatives regarded this training as positive, as an opportunity to interact with other members and to learn, commercial farmers refused to follow the training sessions (they attended only once) because they regarded them as *"a useless time consuming activity"* (Interview CF3, 2010). After the training sessions, not much happened. A committee member explained: *"No more meetings of the WUA took place since over a year now, the secretary is still busy finalizing the constitution or maybe it is somewhere in Pretoria waiting for approval"* (Interview CF3, 2011).

---

[42] It is difficult to estimate the size and composition of the population as no data are available at catchment level. The Community Survey of 2007 estimated the population size of the municipality at approximately 150,000, of which 95% are from previous disadvantaged groups (85% Blacks, 2% Coloureds, 8% Indians) and 5% Whites (STATSA, 2007). The study catchment covers approximately 20% of the municipal area, which brings estimates of the total population in the catchment to 30,000 of whom 28,500 are classified as previously disadvantaged individuals.

[43] 'Emerging farmer' is a policy term in the South African context that refers to historically disadvantaged individuals who are encouraged and supported by the government to develop their agricultural activities for commercial purposes (DA, 2004). Amongst others, these are individuals and communities who benefitted from land reforms and/or are involved in NGO agricultural projects.

4.5 Reflections on inclusion

Taking the existing irrigation boards as a starting point and transforming them into a WUA has given the commercial farmers in the case study area the opportunity to remain fully in command of the process; they could mobilize, reason and take decisions on how to set up the WUA and how to include historically disadvantaged water users. In this process, the commercial farmers made a number of strategic decisions. First, by choosing to establish one WUA instead of four, based on the hydrological boundaries rationale (see also Warner et al., 2008), the four irrigation boards could remain the same: they did not have to include historically disadvantaged individuals within their irrigation boards nor did they have to change their governing rules. In this way, the WUA became a rather empty shell under which the four irrigation boards have continued to operate as they have always done: *"We only formed the WUA because DWA wants us to do so, but in reality the irrigation boards will continue to function as usual, they always managed and will manage the water in this area"* (Interview CF3, 2010).

Second, defining the area under control of the WUA based on the hydrological boundaries of the tertiary catchment and simultaneously stating in the constitution that the objective was to control development in the area gave a powerful tool to protect current water users in the water-stressed catchment. It basically means that initiatives for new water uses, such as for emerging farmers, need to be discussed and agreed upon in the WUA before permit applications can be submitted. Only compulsory licensing as defined in the Water Act can counter this; however, compulsory licensing has not yet taken place anywhere in the country and the instrument is highly contested (van Koppen and Jha, 2005; Movik, 2011). The impact of this strategic decision becomes clear from the responses of the commercial farmers:

> *"I think the WUA will bring many challenges. It will be a challenge to incorporate black people in the management of the institution. But it is very advantageous for us to be able to control the whole catchment and the protection of existing users"* (Interview CF10, 2010).

> *"Users downstream will benefit from the WUA because now the same law applies for everybody and dams to be built upstream or any other activity that may have any repercussion down-stream will be WUA's business...If traditional communities up there want to build a dam they have to ask for our permission now"* (Interview CF6, 2010).

When the scene was set, historically disadvantaged water users were 'included', after which collaborative issues arose such as the language barrier, different comprehension of the role of the WUA and the membership levies that ought to be paid. With English set as the main language, active participation of the Zulu-speaking members was seriously hampered. One of the residents of the former homelands, who is educated as a community development worker, explained: *"some people were afraid to speak at the meetings, they think that they do not know enough and sometimes they do not understand the difficult words used during the meetings"* (Interview F18, 2011). Further, little attention was paid in the first meetings to discuss the reasons for the establishment of the WUA as well as its (potential) roles and organizational arrangements. This made it difficult for the historically disadvantaged individuals to understand the purpose of the meetings nor could they influence the mandate and the structure of the WUA. According to one commercial farmer *"the blacks lost their interests as soon as they realized the WUA would only discuss the management of water and not deal with reallocations of water"* (Interview CF1, 2011). Talks about levies to be raised from the members to make the WUA financially sustainable scared off most of the remaining participants. As expressed by one of them: *"I need good water for drinking and washing...they want me to pay money but they won't solve my problems"* (Interview F9, 2009). The withdrawal of the historically disadvantaged individuals from the WUA demonstrates their agency in defining what inclusion means to them: they refuse to participate in and invest resources in organizations that do not incorporate their interests and in which, in their view, they are forced to accept a subordinate position (see also Cleaver, 1999).

Including historically disadvantaged water users only after the strategic decisions were made, after the rules of the game were set, has had far-reaching consequences for the management of the catchment: the existing domains were left untouched while the new domain of interaction has been captured by commercial farmers, hence, the existing inequities in water control have been reinforced rather than redressed (see also Waalewijn et al., 2005). The success of DWA's role in facilitating the process can be questioned; in fact, facilitation has mainly been left to the commercial farmers who cannot be expected to be impartial in defining the notion of 'inclusion', or indeed be expected to invest considerable amounts of time in facilitating the participatory process required to establish the WUA (see also Brown, 2011).

4.6 Reflection on representation

Reflecting on the composition of the management committee of the WUA (Table 4.2), it becomes clear that the commercial farmers have the largest number of seats compared to the other water user groups, especially when voting rights are taken into account. DWA prescribe a 'balanced representation in terms of the various categories of users' (RSA, 1998) though remain vague about what this means: do they refer to balanced representation of the various water sectors or do the various categories refer to demographic groups? When asked for a response on the imbalanced composition, a regional DWA official responded: *"There are five and five; the four commercial farmers with the Rate Payers Association sum five white members. Then we have two black emerging farmers, plus the Traditional Authority, plus the municipality and KwaZulu-Natal Wildlife Service who are blacks. So that makes five and five"* (Interview O12, 2010). None of the involved parties could explain which water sector the Gender Representative represents and what her role is: perhaps simply meeting the DWA criteria of having 'minority groups' such as women represented in the management committee, or perhaps it is assumed that she will represent domestic water users? Clearly, a view of what is 'balanced' with respect to representation which is limited to skin colour and simplistic gender notions carries the danger of reproducing apartheid philosophies rather than redressing the consequences of them.

Table 4.2: Composition of the WUA management committee.

Emerging sector	Commercial farming sector	Associate members
Emerging farmer Gender representative	Chairperson, Irrigation Board 1 Chairperson, Irrigation Board 2 Chairperson, Irrigation Board 3 Chairperson, Irrigation Board 4	Municipality KwaZulu-Natal Wildlife Service* Traditional authority* Rate Payers' Association**

*Associate members who do not enjoy voting rights.
**Lobby group representing citizens who pay municipal taxes

Securing a seat on a WUA management committee does not automatically mean that the views and interests of historically disadvantaged individuals are represented in the newly established management structures: elements such as authorization, accountability, expertise and resemblance (here defined as the extent to which people feel alike and associated with each other) play a major role in the effectiveness of representation (Brown, 2006). The commercial farmers have a long history of being organized around water; the irrigation boards form platforms which the individual farmers trust to represent their interests at higher levels. Moreover, they have developed a collective identity (Abers, 2007) by framing the problem in terms of biophysical water scarcity and by sharing similar interests, maintaining or preferably increasing their access to and control over water resources to keep their agricultural businesses running. However, in the former homelands located in the study catchment, specific organizations built around water do not exist, since residents there never had the opportunity to use water in large quantities. The existing organizations built around other matters (e.g. land, livestock, crime, marriage) are based on traditional chiefdom structures that have been strongly affected by apartheid and post-apartheid politics (Mamdani, 1996). Consequently, the institutional structures in the former homelands are characterized by fuzziness, with only a limited accountability which has resulted in extensive patronage systems between the traditional leaders, government officials and residents (Kemerink et al., 2011). Authority is primarily based on implicit and competing kinship relations leaving the residents divided. Moreover, as described in detail in Kemerink et al. (2011), the communities are highly diverse in terms of interests, and identifying these residents as farmers merely because they reside in rural areas is a simplistic view of their personal histories and subsequent multiple identities. Resemblance of water-related issues is therefore not straightforward as is reflected in the publicity of the WUA: only three of the 38 interviewees in the former homelands were aware of the existence of the WUA and even the local chiefs did not know who represented them. This shows that the current notion of 'democratic' elected representation applied within the WUA does not make sense from the perspective of historically disadvantaged individuals, since it does not concur with local practices of representation in decision-making processes nor form a strong new platform for interaction on water-related issues to define a common interest.

At the same time there is little understanding from the other stakeholders of why people who are currently not using water in 'relevant' quantities, and who have so far failed to articulate a clear future demand in terms of water use, should be involved (see Waalewijn et al., 2005). As one of the commercial farmers put it:

"To be honest I do not understand why the blacks are in the WUA, they do not use water, so what are we supposed to talk about with them? The black woman from the township who sits in the management committee is growing some tomatoes in a little garden or so. I don't know what she will use more water for, she does not need it" (Interview CF3, 2011)

This inability (or perhaps unwillingness) to adopt a forward looking view of the future, in which emerging sectors require larger quantities of water, is supported by a local DWA official who stated that *"representation in the management committee has to be directly related to land ownership"* (Interview O2, 2010); he argued that stakeholders who own more land and water should have a bigger say in decision-making as their stakes are higher. This reasoning implies that inequity in access to land legitimizes inequity in access to decision-making platforms. This is in direct contradiction to the progressive stand on representation that DWA defines in its official policy documents, in which it is realized that transformation of the water sector will not take place if the current possessors dominate the new water organizations. Discrepancy on the interpretation of how representation ought to be defined between the national and local DWA officials carries the risk of jeopardizing the reform process which is aimed at under the Water Act.

4.7 Discussion

This paper shows that the establishment of the new institution has been unsuccessful in contributing to transformation in the case study area. On the contrary, the WUA is currently a sleeping giant, though when it is fully awake it will potentially steer to benefit the haves over the have-nots:

> *"WUA is a way for us as commercial farmers to obtain licenses to construct new dams now based on inclusive grounds...I would be happy if we could build a dam up in the tribal lands, we can pay for it and we give them a share...though I can already tell you, they will not use the water, so we will end up using their share as well"* (Interview CF4, 2010)

As argued by Von-Benda-Beckmann and Von-Benda-Beckmann (2006), laws such as the National Water Act get renegotiated, interpreted and rearranged at local level and the outcomes tend to reflect the existing power relations within society. Surely one should question why and by whom the existing structures were chosen as points of departure for the implementation of the Water Act, as they have further reinforced existing inequities. This underpinning of the inequities is not only reflected by the decision to allow the irrigation boards to be in charge of establishing the WUAs, but also in the way that existing water use was recognized as lawful in the study catchment (see Movik, 2011). Without proper studies being made of the water availability in the catchment and in the absence of enforced monitoring of actual use, commercial farmers were simply asked to register their water use. This gave them the opportunity to register additional water use in anticipation of future use and/or reallocations (Méndez, 2010; Méndez et al., *forthcoming*). But is what we see happening within the case study area solely the result of the implementation process, in which negotiation, interpretations and rearrangements have taken place at different spatial levels, or are there more fundamental issues at stake? Are the concepts that have been chosen within the water reform processes, such as WUAs, the right vehicles to achieve the transformation of the water sector as envisaged in the South African Water Act?

The concept of WUAs is a prime product of institutional crafting theory. The idea that institutions can be crafted (Ostrom, 1992) is widely embraced by governments and development agencies who prefer working with institutional blueprints (Roe, 1991; Mosse, 2004; Rap, 2006; Molle, 2008), though widely criticized in academic circles (Giddens, 1984; Long and van der Ploeg, 1989; Cleaver, 2002; Boelens, 2008; Molle, 2008; Ahlers, 2010; Laube, 2010). Policies based on institutional crafting are dominated by rational choice thinking, in which it is assumed that individuals make the appropriate calculations of costs and benefits based on single preferences, leading to an inclination for clearly visible, democratic, legally recognized institutions for participation. Within these kinds of institutions, the notion of inclusion is closely related to those who are visible, for instance through actual use of water, and those who are recognized by the others to hold some kind of authority, for instance on the basis of the possession of (natural) resources or particular knowledge. Representation is often based on clearly delineated stakeholder groups who find resemblance by sharing a collective identity and who democratically elect a representative that they can hold accountable (Brown, 2006). These notions on inclusion and representation originate from a neoliberal inclined normative order that propagates particular views on essential issues, such as the relationship between the individual and the community as well as independency versus dependency, which do not always match with other normative orders upheld in society (see Wolf, 2008). The analytical framework of legal pluralism that has been adopted in this paper recognizes the various normative orders in society and provides contextualized insight into the diverse notions of inclusion and representation that delineate specific institutional inclinations. This is essential for understanding the dynamics of institutional evolution, including the contested space that has emerged with the introduction of WUAs as new platforms for interactions over water.

The history of apartheid has resulted in two separate worlds in one country (Bond, 2007; Cousins, 2007) with distinctive normative frameworks. The South African commercial farmers have been brought up within a similar neoliberal normative standpoint as the institutional crafting theorists who developed the concept of WUAs (de Lange, 2004). Hence, the set-up of the WUA with its explicit organizational structure and the focus on functional and managerial issues is socially embedded within the commercial farmers' community. They have extensive experience managing water within similar organizations and have built a strong collective identity. Thus, even though they did not see the need for it, the commercial farmers were easily brought on board, as they knew the rules and they knew how to bend them. However, the WUA set-up is not in line with the normative orders prevailing within the former homelands where different institutions and practices prevail that are more implicit and based on kinship. Democratically elected representation is not fully recognized, water use is considered insignificant and a collective identity around water does not yet exist. This has left stakeholders within the former homelands invisible and without a voice. In other words, it is difficult for the people residing in the former homelands to be effectively included and represented in the way defined by the current set-up of the WUAs. This demonstrates that the notions of inclusion and representation entrenched in the WUA concept are far more biased than is acknowledged by policy makers. Moreover, by adopting neo-liberal inclined institutional blueprints, existing inequities are legitimized and hence further strengthened within the new water institutions. To evade this, a more profound process which deals with historic inequities needs to take place without victimizing or ignoring the multiple social identities of all the actors, and taking into account the various normative orders that exist in society. For collaboration in water management, this means opening up space for bargaining not only over the content but also over the institutions that govern this collaboration. Unless this inherently political participatory process is initiated and the different institutional preferences become part of the negotiation process for the 'why' and the 'how' of progressive collaboration at catchment level, the establishment of the WUA in the study catchment will not contribute to achieving the transformation envisioned for rural South Africa.

5. Why infrastructure still matters: unravelling water reform processes in an uneven waterscape in rural Kenya [44]

Abstract

Since the 1980s, a major change took place in public policies for water resources management. Whereas before governments primarily invested in the development, operation and maintenance of water infrastructure and were mainly concerned with the distribution of water, in the new approach they mainly focus on managing water resources systems by stipulating general frameworks for water allocation. This paper studies the rationales used to justify the water reform process in Kenya and discusses how and to what extent these rationales apply to different groups of water users within Likii catchment in the central part of the country. Adopting a socio-nature perspective, this paper shows how the water resource configurations in the catchment are constituted by the interplay between a normative policy model introduced in a plural institutional context and the disparate infrastructural options available to water users as result of historically produced uneven social relations. We argue that, to progressively redress the colonial legacy, direct investments in infrastructure for marginalized water users and targeting the actual (re)distribution of water to the users might be more effective than focusing exclusively on institutional reforms.

---

[44] This chapter is based on: J.S. Kemerink, S.N. Munyao, K. Schwartz, R. Ahlers, P. van der Zaag (*forthcoming*) Why infrastructure still matters: unravelling water reform processes in an uneven waterscape in rural Kenya. Under review *International Journal of the Commons*.

5.1 Introduction

Since the 1980s a major change took place in public policies for water resources management. The general objective of policies shifted from an emphasis on physical water delivery by governments to creating an enabling environment for other parties to provide water services. Whereas before governments primarily invested in the development and operation of hydraulic infrastructure and were mainly concerned with the distribution of water, in the new approach they mainly focus on managing water sector by stipulating rules for water allocation (Cleaver and Elson, 1995; Allan, 1999; Mosse, 2004; Lowndes, 2005; Swatuk, 2008; Saletha and Dinar, 2005; Mosse, 2006; Ahlers and Zwarteveen, 2009; Sehring, 2009). Based on this global shift in public policy approach the Kenyan government revised its water legislation in the early 2000s. They took up primarily an oversight role in the water sector in which they attempt to steer and control institutions that govern decision making over water resources by drafting regulatory frameworks, disseminating organizational blueprints and specifying key principles. As such, rather than directly manipulating the distribution of water resources through investments in infrastructural development, the Kenyan bureaucrats became involved in crafting an institutional change process in the hope that it would lead to specific material outcomes aligned with their political ideals and ambitions envisioned in the policy reform process. But how does this shift in policy approach unfold in practice and how does it affect water resource configurations, here defined as the materialized division in control over, access to, and distribution of water between water users sharing the same water resource?

Reforms in public policies are based on a certain logic, a rationale that articulates an assumed 'common sense' and which is used to justify the policy intervention. This paper studies the logics used within the ongoing water reform process in Kenya and discusses how, and to what extent, these logics apply for different groups of agricultural water users within a Kenyan waterscape. Based on this, we will show how the current reform process produces inequitable as well as paradoxical outcomes and we will argue that the current partial focus on institutional change processes, ignoring physical aspects such as access to infrastructure and diverse geographical conditions in which water is being used, leads to these specific outcomes. In this way the paper will contribute to the broader discussion on the implications of water reform processes on the management and use of this common pool resource (Cleaver and De Koning, 2015).

The Likii catchment, located on the slopes of Mount Kenya in the central part of the country, is used as a case study. Under the water reform process that commenced in the late 1990s, nine water user associations (WUAs) have been established within its sub-catchments and one overarching river basin water user association (RWUA) at catchment level. The research focused on the nine WUAs within the catchment as well as the RWUA. The findings presented are based on in-depth semi-structured interviews with thirty-five small-scale farmers and four large-scale water users within the catchment carried out between October 2010 and January 2011. The interviewees were selected by a stratified random selection procedure to guarantee geographical spread and to obtain input from various age, ethnic, class and gender groups. The findings of the interviews were cross checked through focus group discussions, observations, comparison with existing literature and by consultations with other actors such as local authorities, government officials and non-governmental organizations (NGOs) active in the region.

This paper first explores the theoretical considerations used to analyze the water reform process. In the next section a detailed narrative of the Likii catchment is provided including a historic analysis of the water resource configurations. Thereafter we describe the 2002 water reform process as stated on paper and how it unfolded within the case study catchment. This is followed by a critical reflection on the logics used to justify the water reform process as well as the implications on control over, access to and distribution of water for various water user groups. In the concluding section we analyze the interplay between geographical conditions, technical options and socio-political arrangements that constituted the water resource configurations in the waterscape and reflect on the implications for policy makers concerned with equity issues and aiming at inclusive development.

5.2 Theoretical considerations

In this paper we adopt a critical institutionalist's perspective in which we conceptualize institutions that govern water resources as outcomes of dynamic social processes in which authority is constantly contested, negotiated and reaffirmed (Cleaver and De Koning, 2015). In the constant reordering of environments, unequal social relations play a central role "in determining how nature is transformed: who exploits resources, under which regimes and with what outcomes for both social fabrics and physical landscapes" (Budds, 2008:60; see also Leach et al., 1999; O'Reilly et al., 2009). Similarly, Swyngedouw states that "the mobilization of water for different uses in different places is a conflict-ridden process and each techno-social system for organizing the flow and transformation of water (through dams, canals, pipes, and the like) shows how social power is distributed in a given society" (Swyngedouw, 2009: 57). In this process not only the agency of social actors play a role in forming dynamic waterscapes, but also the agency of the physical environment. Ecological relations shape and reshape societies and circumscribe the ever changing range of choices available for human exploitation. Moreover, once constructed, hydraulic infrastructure is not merely a passive object, but a force in itself, capable of rearranging and affecting water flows and as such constitutive of authority as it opens and forecloses certain decisions and future trajectories (Ahlers et al., 2011; Meehan, 2014; Van der Kooij et al., 2015). Water resource configurations can therefore be conceptualized as outcomes of a mutually constituted interplay between geographical conditions, available technologies and socio-political arrangements (Swyngedouw, 2009; Mosse, 2008). Swyngedouw argues that understanding water resource configurations as historically produced rather than based on logic calls for "a transformation in the way in which water policies are thought about, formulated and implemented" (Swyngedouw, 2009:56).

To understand the contemporary policy making process, this paper uses an analytical frame that emphasizes the political nature of policies. In this framework policies are regarded as outcomes of a discursive practice of policy making in which problems are framed and ideas, concepts and categories are aggregated through which meaning is given to a particular phenomenon (Hajer, 1995; Mollinga 2001; Griggs, 2007). Several scholars argue that specific storylines referred to as policy narratives are influential within the policy-making process (Roe, 1994; Hajer, 1995; Mosse, 2004; Rap, 2006; Molle, 2008; Peck and Theodore, 2010). These policy narratives, the discursive form in which a particular logic is presented, can be understood as specific and stabilized interpretations of physical and/or social phenomena that assume certain causal relationships not necessarily grounded in empirical evidence (Roe, 1994; Molle, 2008). The persistence of policy narratives, even in the absence of empirical grounds, can be seen as the result of the continuous support from actors active within policy networks. These epistemic communities or expert networks gradually get established within the process of the proliferation of a policy in which actors share ideological understandings and cultural practices (Conca, 2006; Rap, 2006; Molle, 2008; Peck and Theodore, 2010). Policy narratives form part of the larger theoretical story of how the network understands reality, based on their ideologies and interests, and as such the members of the network have an incentive to maintain particular policy narratives (Rap, 2006; Mosse, 2004).

The policy narratives produce and legitimize certain policy models, a prescribed set of principles, procedures and structures that together provide a 'blueprint' for intervention to address a particular issue in different locations (Rap, 2006; Molle 2008; Peck and Theodore, 2010; Rusca and Schwartz, 2012). In this way, a policy model obscures its ideological origin and is often disconnected from local realities, possibly producing different outcomes in different contexts. Policy models are widely embraced by governments and development agencies. Policy models fit well with the positivist aims for 'objectivity' and 'neutrality' that are dominant within the development orthodoxy as it assumes that performance of the standardised policy can be measured and compared based on predefined indicators (Power, 2000; Rap, 2006; Peck and Theodore, 2010). Not only does this ease the work procedures of government agencies; adopting policy models also conveniently depoliticizes the policy making process (Mosse, 2004; Molle, 2008). Conca argues that policy networks with particular value orientations, through circulation of information, framing problems and solutions, and pressuring governments, have become an *"authoritative source of norms in world politics"* (Conca, 2006:126; see also Goldman, 2007; Peck and Theodore, 2010). National governments in the global south are often pressured by these policy networks to conform and adopt similar policy models in order to receive legitimacy and possibly financial support (DiMaggio and Powell, 1983; Lodge and Wegrich, 2005).

It is within this theoretical understanding that we analyze the policies that are propagated within the water sector reforms in Kenya. We acknowledge that policy making is a highly dynamic process and at any point in time several (overlapping) policy networks may exist at different spatial levels (see also Funder and Marani, 2015). These policy networks might have different normative views and aim to pursue different interests within the same policy domain and as such compete for authority. After all *"... hegemony ... is an always incomplete process. The powers of network-normativity and model-making may be formidable, but they are far from totalizing, since they are also marked by contradiction and contestation"* (Peck and Theodore, 2010:171). This contested process may lead to changes in the content of policies as well as to differences in policies at various locations. Nevertheless, within the reform processes ongoing in Kenya we observe striking similarities with narratives used to justify the reform processes elsewhere that lead to the implementation of similar policy models in dissimilar contexts (Cleaver and Elson, 1995; Rap, 2006; Ahlers and Zwarteveen, 2009; Manzungu and Machiridza, 2009; Sehring, 2009; Kemerink et al., 2011; Mtisi, 2011; Manzungu, 2012; Van Koppen et al., 2014). It is therefore crucial to scrutinize these narratives and the policy models they promote and legitimize, and to illuminate how they unfold within the historically produced uneven waterscapes such as the Likii river catchment.

5.3 Setting the Scene

Likii river catchment is located on the north-western slopes of Mount Kenya within the Upper Ewaso Ngíro North Basin (see also Figure 5.1). The catchment has an area of 174 km^2 with altitudes ranging from about 5,000 m above sea level in the upper parts of the catchment to about 2,000 m above sea level in the lower part. The upper part has a cool, wet climate with a mean rainfall of 1,100 mm/year and is covered by forest, bush land and grassland on deep soils. The lower part has a semi-arid climate with a mean rainfall of 750 mm/year and is covered with savannah vegetation on alluvial soils (Kiteme and Gikonyo, 2002; Rural Focus, 2009). The catchment has two dry seasons per year, from January to March and September to November, causing low flows in the main Likii River and with some tributaries drying up completely.

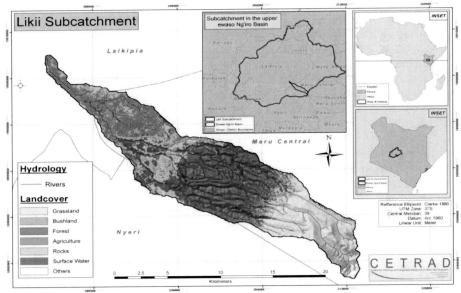

Figure 5.1: A nested map of Likii catchment with its position in the Upper Ewaso Ngíro North Basin, Kenya and Africa (CETRAD, 2010)

At the onset of the British colonial administration in Kenya in the early 1900s, white settlers migrated to the country lured by the prospect of land. They settled in the fertile highlands outside Nairobi including the Likii catchment, dispossessing the Maasai and the Kikuyu tribes from their ancestral lands. The farmers abstracted water from Likii River through diversion channels (furrows) and, with government support, they sunk boreholes and constructed dams as supplemental sources of water. Under the Water Ordinance of 1929, and the revision in 1951, most settlers obtained (provisional) water rights for their water abstractions (Nilsson and Nyanchaga, 2009). In 1963, Kenya acquired independence and the new government revised the legal structures. In response to this transition most white farmers left the mountain highlands and sold the land to investment companies. These companies sold the land in 0.8 hectare plots to subsistence farmers from the neighbouring Nyeri and Meru districts. Upon dividing the land, the water permits of the white farmers were declared redundant by the Water Apportionment Board and the water rights were therefore obsolete by the time the new water users moved in. Nevertheless, two settlers' farms were bought up, including the water rights and hydraulic infrastructure, by foreign companies and turned into commercial flower farms.

Currently, circa 60,000 people reside in the Likii catchment with the highest population density in the midlands (Rural Focus, 2009). The highlands are part of the Mount Kenya Wildlife Conservancy and form the source of water for the utility supplying water to the nearby towns. The livelihood of the majority of people in Likii river catchment largely depends on small-scale subsistence agriculture with few people being employed at the commercial flower farms located in the middle and lower reaches of the catchment. In the lower part of the catchment cattle is herded in addition to subsistence farming. The rainfall variation in the catchment often has a detrimental effect on crop production. As a strategy to cope with uncertainty and poor distribution of rainfall during the cropping seasons, the farmers have constructed furrows to abstract water from the river for supplementary irrigation (Rural Focus, 2009; see also Rajabu and Mahoo, 2008). The small-scale farmers typically collaborated in the construction and maintenance of the furrows (see for similar examples in Kenya Fleuret, 1985, and in Tanzania Kemerink et al., 2009; Komakech et al., 2012a; Komakech et al., 2012b). This organization around water was partly an internal process based on the need for collective action to access water and partly assisted by external parties such as NGOs and relief agencies. Not all farmers joined, some farmers had land close to the river so they could easily access water independently while others remained dependent on rainfed agriculture. Since the late 1970s, many water user groups were registered as so-called self-help groups with the Ministry of Culture and Social Services (see also Table 5.1) and applied for provisional water rights under the Water Ordinance of 1951. These provisional rights would enable these groups to construct the necessary water works, often funded by (foreign) NGOs with labour provided by the farmers. The formal water permit would only be issued after the completion of the works, which included intake structure, storage reservoir to ensure uninterrupted supply of water during the dry seasons, field canals, and installation of measuring and control devices. In the case study catchment these constructions resulted in irrigation systems with centralized storage tanks of about 200 m^3 from which farmers receive water through a 0.5 inch pipe with a discharge of about 43 m^3 per day at full pressure based on design parameters (see also Figure 5.2). However, the actual discharges are considerably lower; the tanks only serve as overnight storage due to their limited capacity, leading to water shortage during dry spells, and additional pipes have been added after the design of the irrigation systems. Moreover, the discharge differs considerably per irrigation system as in three out of the nine irrigation systems no storage tanks have been installed. The four large scale users in the catchment, the two flower farms, the water utility and the wildlife conservancy, abstract together circa 46% of the total amount of water abstracted from the Likii River (Rural focus, 2004).

Table 5.1: Details of the nine water user groups established in Likii catchment as well as the large scale water users (source: WRMA archives)

Name Water User Group	Year established	Initial members	Current members	Estimated command area (Ha)	Average plot size per user (Ha)	Year of provisional water right
Mukuria	1978	190	500	320	0.6	1979
Miarage A	1979	160	120	320	2.0	1987
Nkando	2002	200	200	162	0.8	2010
Nturukuma	1987	150	640	1700	2.4	1991
Mukima	1992	50	158	2000	12.1	2004
Miarage B	2000	250	300	Unknown	Unknown	2004
Jikaze	2001	39	38	32	Unknown	Not yet
Murimi	1999	105	280	Unknown	Unknown	2004
Kiranga	1992	190	326	360	1.0	1993
Large scale users						
Flower farm 1	1989	1	1	37	36.8	1994
Flower farm 2	2001	1	1	26	25.9	2002
Mt Kenya Wildlife Conservancy	2000	1	1	223	223.0	2003
Water Utility	2008	1	1	n/a	n/a	Not yet

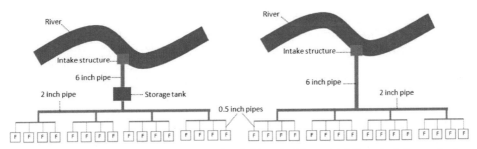

Figure 5.2: Sketch irrigation systems with (left) and without (right) central storage tank.

5.4 Narrating the Kenyan water reform process

It is within the context described above that the water sector reform process took place through the implementation of the National Water Policy of 1999 followed by the National Water Act in 2002 (Republic of Kenya 1999; 2002). Similar to water reforms that took place elsewhere in the world since the 1980s, the Kenyan water reform policies no longer emphasize the role of the government to provide water to the citizens, but focus on creating an enabling environment for other parties to provide water. As such the Kenyan government, technically and financially supported by the World Bank and overseas development agencies, reduced its investments in the development, operation and maintenance of water infrastructure and concentrated its efforts on managing water resources (Sambu, 2011). Three prevailing narratives can be identified within the official Kenyan policy documents, which resemble the dominant narratives used to justify water reforms in other countries (see also Cleaver and Elson, 1995; Savenije and Van der Zaag, 2002; 2008; Van der Zaag, 2005; Conca, 2006; Molle, 2008; Anderson et al., 2008; Ahlers and Zwarteveen, 2009).

The first narrative relates to the assumption that lack of entrepreneurship, such as private investments in hydraulic property and commercial use of water, are caused by the absence of, or insecurity in, tenure (Cousins, 2007; Molle, 2008; Peters, 2009). Incorporating this narrative the Kenyan Water Act calls for a revision of the water right regime to provide security in water use for private entities. Building further on the previous legislation, under the Water Act ownership of water exclusively belongs to the State and all uses of water, except for domestic purposes, will be bound by conditional permits (Republic of Kenya, 2002: articles 3, 26 and 25; see also Sambu, 2011). Permits can be obtained by individuals and legally recognized private entities through administrative procedures that vary based on the amount of water requested. For issuing a permit and determining any conditions to be imposed on a permit, amongst others, the existing lawful uses in the catchment, the efficient use of the water and the strategic importance of the proposed water use are taken into account (Republic of Kenya, 2002: article 32). The Act only provides narrow room for rejecting permit requests based on inequity in allocation; only in case of changing (environmental) conditions (Republic of Kenya, 2002: article 35a) and/or in specific geographical areas (Republic of Kenya, 2002: article 36b).

The second narrative is based on the hypothesis that centralized decision making leads to decisions that are not sufficiently aligned with the interests and context of actors at local level and that these decisions are therefore not effective. It is assumed that when decision-making is decentralized, local actors have better access to decision-making platforms to participate, monitor and/or use pressure on those involved in decision-making processes. In this way, local actors are believed to have greater influence on decisions that affect them and as such the outcomes would fit better with local realities (Cleaver, 1999; Goldin, 2013; Kemerink et al., 2013). Based on this rationale the Water Act stipulates the need for public consultation within the application process for water permits (Republic of Kenya, 2002: article 107) especially to strengthen the control of water use by private entities (Republic of Kenya, 2007). As such, the Water Act effectively sets the stage for the decentralization of water management responsibilities to newly established authorities at lower administrative levels that will facilitate stakeholder participation within new platforms at catchment level as well as formalization of WUAs at sub-catchment level (see also Kiteme and Gikonyo, 2002; K'Akuma, 2008; Robinson et al., 2010).

The third and last narrative links low prices with wasteful use. It argues that if goods or services come for free or at a low price, users tend to waste it, while if it comes at a higher price they will use it more efficiently (Molle and Berkoff, 2007). Within this narrative setting 'correct' prices is seen as a tool that facilitates optimal allocation of scarce resources, in this case water, among competing uses and stimulates users to enhance efficient use of the resource (Johansson, 2000). Incorporating this narrative the Kenya government introduced the payment of water use charges as part of the Water Act (Republic of Kenya, 2002: articles 31 and 60) as tools for demand management and stimulating social and economic use of water (Republic of Kenya, 2007: article 105-2c and 2d). The income generated with charging fees for water use is meant to recover the actual costs of *"managing the water resources and water catchment areas"* (Republic of Kenya, 2007: article 105-2a). For each river basin this has resulted in specific tariffs that depend on the available water in the basin, the amount of water requested and the purpose of the water use. According to the Act water used for domestic and subsistence purposes is free of charge.

5.5 The Kenyan policy model

Each of the narratives discussed in the previous section called for particular interventions that together pave the way for rolling out a comprehensive policy model to facilitate the reform process. This section will describe the policy model as has been designed and implemented within the reform process in Kenya.

As a central part of the water reforms and under the direct auspices of the Ministry of Water, the Water Resources Management Authority (WRMA) was established at national level with offices at regional and sub-regional (local) level to assist the implementation of the Act. The Authority has the overall responsibility for the management of the water resources and is in charge of approving water permits applications and charging water use fees. One of the main roles of the WRMA is to ensure stakeholder involvement by initiating and facilitating the establishment and formalization of river basin water user associations (RWUAs) at catchment level. According to the Water Act, the RWUAs will advice on issues concerning water resources conservation, the use and apportionment of water in catchment areas and will consist of representatives of the (local) government agencies and representatives of business communities, farmers, pastoralists and NGOs (Republic of Kenya, 2002: article 16.3).

To protect the interests of the existing water user groups who already shared hydraulic infrastructures, the Water Act introduced WUAs at local level. Within the Water Act the WUAs fall under the definition of 'community projects' which are projects authorized by WRMA and operating under a permit to use water or to drain land that is endorsed by at least two-thirds of the persons owning or occupying the particular project area (Republic of Kenya, 2002: articles 19 and 23). As such the WUAs are legally recognized and regarded as private entities which make WUAs eligible to apply for water permits as well as to be represented in the RWUA.

Under the 2002 Water Act, the old procedure of granting water permits has been revised in order to better scrutinize the new applications and subject them to wider stakeholder consultations. Permit applications have been classified into categories A, B, C and D for both surface and ground water abstractions depending on the severity of impact the water use activity is perceived to have on the water resources (see Table 5.2). All permits are issued usually for five years after approval at the appropriate level (local, regional and national). A copy of every water permit application should be submitted for comment to the relevant registered RWUA, if one exists, *"to ensure that the proposed water use does not affect other users in any way"* (Republic of Kenya, 2007: 28). The RWUA is expected to recommend the application after inspecting if the water requested is available at the intake and downstream users are not negatively affected. Otherwise the RWUA is expected to consult with the applicant and the water users to agree on modalities of minimizing or mitigating the potential effects of the proposed water use (Kiteme and Gikonyo, 2002). After the application is reviewed and approved by the WRMA a provisional permit is given which allows the applicant to start constructing the appropriate water works, including measuring and control devices and a 90-day storage facility, within a certain timeframe. After completion, the works are inspected by the WRMA and if deemed satisfactory the permit is granted.

Table 5.2: Category system for granting water permits and charging fees as applied in the Upper Ewaso Ngiro North Basin (Republic of Kenya, 2007) *(exchange rate August 2011: 1000 Ksh= 7.50 euro)*

Category	Description	Water Quantity (m³/day)	Permit fees (Ksh)		Procedure		Water charges (Ksh/m³)	
			Assessment of application	Renewal of permit	Recommends	Approval authority	Domestic/ Livestock	Irrigation/ Commercial
A	Water use activity deemed by virtue of its scale to have a low risk of impacting the water resource	< 50	1000	Nil	RWUA	Local	Nil	Nil
B	Water use activity deemed by virtue of its scale to have the potential to make a significant impact on the water resource	50 - 500	5000	7500	RWUA	Regional	0.5	0.5 < 300 m³/day 0.75 > 300 m³/day
C	Water use activity deemed by virtue of its scale to have a significant impact on the water resources	500 - 5000	20000	25000	RWUA	Regional	0.5	0.5 < 300 m³/day 0.75 > 300 m³/day
D	Water use activity which involves either two different catchment areas, or is of a large scale or complexity and which is deemed by virtue of its scale to have a measurable impact on the water resource	> 5000	40000	50000	RWUA	National	0.5	0.5 < 300 m³/day 0.75 > 300 m³/day

Once the permit is granted the permit holder needs to pay water fees depending on the amount of water that they abstract (see Table 2). To protect small-scale water users, Category A users with water abstraction of less than 50 m³ per day are not charged any water fees. The tariff setting in category B, C and D is partially progressive for water used for agricultural and commercial purposes: two blocks are defined with 300 m³ per day as the threshold. Under the new laws the water fees also became applicable for existing permit and authorization holders granted under the previous Water Act.

5.6 Unfolding the policy model in Likii catchment

In 2001, a process was initiated to establish the overarching Likii river basin association (RWUA) in line with the legal reforms and in response to perceived 'imminent conflicts' over water between upstream and downstream users. This process was initiated by the Laikipia Research Programme funded by the Swiss Development Cooperation together with one of the large-scale flower farmers who had the capability and resources to mobilize the parties involved. The initiators decided that membership of the RWUA was confined to the nine WUAs and the large scale water users i.e. flower farmers, water utility, nature conservation and tourism enterprises. Other users, such as small-scale individual irrigators and pastoralists, were thus excluded beforehand from a voice in the RWUA. The Likii RWUA was registered as an association by the office of the Attorney General in 2002 based on the payment of a registration fee, the RWUA constitution, membership list, and the minutes of election meetings of the office bearers. The contents of the constitution had to a large extent been copied from a neighbouring RWUA established earlier. The initial mandate of the RWUA was to oversee good water resources management at the river catchment level, including coordinating water abstractions and conflict mitigation. However, the RWUA later expanded its mandate to address issues of water pollution and riparian land degradation, as well as water use efficiency. For water allocations, the Likii River was divided into four sections, each dedicated to serve particular groups of water users (see also Table 5.3) with specific allocation schedules during rationing periods. Since its establishment, the RWUA plays an important role in monitoring the water abstractions and ensuring that water users stick to their water turns. The large-scale users have self-regulating intake devices, whereas for the WUAs all intakes have been equipped with control valves enclosed in lockable chambers which have three locks each for the WUA, Likii RWUA, and local office of the WRMA respectively. This means that during the opening or closing of the control valve during the rationing periods all three parties need to be present and other parties cannot purposely close the gate valves during the water intake.

The positions within the RWUA management committee are also allocated based on the river sections and the number of water users they represent (see also Table 3). It should be noted that the committee members appointed from the Likii North and Likii main river sections include the owners of the two commercial flower farms. The small-scale farmers are indirectly represented in the management committee by five out of the nine chairpersons of the WUAs. The large-scale farmer who initiated the process holds the powerful position of secretary and in line with the constitution to ensure impartiality, the first chairperson of the RWUA management committee was an external person. However, the person selected was a commercial farmer from a neighbouring catchment who, according to an interviewee, maintained close relations with the secretary of the RWUA (Interview RL7, 2010).

Table 5.3: Subdivision of the Likii river with mean flows, users and initial division of positions within the RWUA management committee including the ratio between positions occupied by representatives from the commercial farmers (c) and representatives from the small-scale farmers (s) (WRMA archives, 2011)

River section	Mean river flow (m³/s)	Large-scale users	# WUAs	# Users represented	# Management positions allocated (c/s)
Likii north	0.593	Flower Farm	5	873	3 (1/2)
Likii central	0.479	Public Water Utility	0	1	1
Likii south	0.328	Mt Kenya Wildlife Conservancy Mt Kenya Safari Club	0	2	2
Likii main	0.140	Flower Farm	4	337	4 (1/3)

The 2002 Water Act offers the RWUAs an important legal instrument to control access to water in their catchment and to direct developments by involving them in the permit application process. The Likii RWUA has set the following criteria in its constitution for giving a positive recommendation to the WRMA on a proposed abstraction:

1. The amount of water at the proposed source meets the applicant's demand
2. The abstraction will not negatively affect existing downstream entitlements
3. In case of a small-scale user: no WUA serves the residential area of the applicant
4. In case of a small-scale user: the water originates from another source than the river (e.g. a spring, borehole or dam).

When criteria 3 and 4 are not fulfilled, the permit application will not be recommended to the WRMA. These extra criteria set by the Likii RWUA basically forces small-scale farmers to join the WUAs in case they want to abstract surface water from the river. Only water use rights from springs, dams and boreholes can be acquired on individual basis by small-scale users in contrast to large-scale users, whose individual applications of river abstractions will be considered for recommendation to the WRMA. Even though it was not in the formal policies, the Ministry of Water supported the initiative as according to a Ministry official: "it is easier to administer water rights through groups than dealing with these individual small-scale water users directly" (Interview S5, 2010, see also Funder and Marani, 2015). Since the introduction of the required recommendation by the RWUAs in 2007, the Likii RWUA has recommended four abstractions from springs and two from boreholes. In addition, three abstractions from springs were rejected as "they tapped all the water from springs flowing into the Likii River, leaving no flow for downstream use" (Interview RL7, 2010). No applications for abstractions directly from the river have been submitted to the RWUA so far.

As part of the reform process the existing water user groups became formally recognized by the WRMA as so-called community projects, generally referred to as WUAs, and as such are the main formal bodies in which small-scale farmers are organized. Only through the structure of the WUAs small-scale farmers can be represented within the RWUA, which extended the mandate of WUA management committees as well as their executive powers. As shown in Table 5.1 the number of members in some WUAs has increased considerably since the establishment while other WUAs remained the same or even decreased in number. In-depth analyses of two upstream and two downstream located WUAs show other significant differences in the management of the WUAs. According to the constitutions, elections are organized every two years. However, in two WUAs the office bearers change frequently allegedly as result of incompetence and disinterest in organizational matters, while in other WUAs few changes are made in leadership positions. In one WUA interviewees report that office bearers receive water through 1.0 inch pipes while to the other users the water is supplied through smaller pipes (Interview F29, 2010), a material benefit that potentially explains the reason to cling to power in some WUAs. In another WUA members reported allegations of misappropriation of WUA funds (Interviews S1 and W3L3, 2010). No reservations are made in the constitution on the number of terms that office bearers can be elected, so they can stay in their position as long as they get re-elected. Despite democratic ambitions, kinship and patronage systems seem to determine the appointment of office bearers within the WUAs, with positions circulating within small a group of (rival) local elites. Also the membership fee to join the WUAs varies considerably, ranging from Ksh 29,000 up to Ksh 100,000 per connection[45]. In addition members are expected to pay seasonal fees for construction, operation and maintenance of the infrastructure. However, it remains unclear how the incoming money has been spent by the WUAs. So far, none of the WUAs have complied with the Water Act to construct a 90-day storage facility nor have they enlarged the intake structure and piped system to accommodate the increased number of members. Even though the WUAs did not fulfil the legal criteria to receive the water permit as they did not yet construct the required hydraulic storage infrastructure, the WRMA already granted the permit for three WUAs and is processing the permits for the others. As a result, the WUAs have started paying for the water fees to WRMA as given in Table 2. Although the individual water use of the farmers is below the threshold of 50 m^3 per day, which would allow them to abstract for free, their collective water use required them to obtain a permit in category B or in some cases even C.

5.7 Unravelling the implications for water users

In this section we will analyze the implications of the water reform process for three groups of water users: the large-scale commercial farmers, the small-scale users who are members of one of the nine WUAs, and the small-scale users who are not member of a WUA. For each group of water users we will validate the three narratives that were used to justify the reform process and reflect on what the selected policy model meant for them including the implications on their access to water.

For the large-scale commercial farmers the reforms meant an increase in the price they had to pay for water. To what extent this has affected their business remains unclear, though their modern infrastructure such as self-regulating devices allows them to abstract and store the exact amounts they require and limits the water loss in the system. This gives them the opportunity to find the economic optimum for their agricultural business and the commercial farmers irrigate both during the rainy and dry season with storage capacities up to 160,000 m^3. Their privilege of holding water use rights on an individual basis means they have secured access to water and they can apply for renewal or an amendment on the volume of water on their own initiative. Moreover, the reforms allowed the large-scale users to further strengthen their control over the water developments within the Likii catchment. They are individually represented in the RWUA and the large-scale users together occupy a disproportional number of positions in the management committee (see also Table 5.3). This means they can directly

---

[45] Based on the exchange rate of August 2011 this equals 220 to 750 euro. In comparison, the GDP per capita in 2011 has been estimated at 1,255 euro (CIA, 2012)

influence which water abstraction applications get recommended to the WRMA and which ones do not get recommended. Even though most WUAs already had (provisional) water use rights before the RWUA got established, any application for a permit renewal or an amendment of the abstracted amount of water must be submitted to the RWUA for comment, which to some extent protects the access to water for the large-scale farmers in the longer term. For the large-scale commercial farmers the three narratives that were used to justify the reform process seem to apply (or at least not proven invalid by this research) and the adopted policy model under the water reform process has had an overall positive effect on their (future) access to water.

Even though the aim was to increase the water security for small-scale farmers through formalizing existing community organizations into WUAs and recognizing WUAs as private entities eligible for water permit applications, it worked out differently for most members of the nine WUAs in the Likii catchment. The Likii RWUA basically forces all small-scale farmers to join WUAs, which has increased considerably the number of members in six of the nine WUAs. Without sufficient storage in the system and without an amendment of the permit to abstract more water, it means less water is available per person: on average the farmers are currently able to only supplementary irrigate 0.1 hectare during the rainy season. During the dry season only 24% of the farmers are able to irrigate, mainly those who privately constructed small reservoirs or who have plots located at hydraulically advantageous positions within the system (Munyao, 2011). Consequently, most farmers are seriously constrained to enlarge their agricultural production due to water shortage and in addition are dependent on a time consuming and complex communal process to improve the system. Yet at the same time the costs for accessing water have increased considerably: they are obliged to pay water fees in a higher water fee category and in some WUAs they are charged considerable membership fees to get connected. In other words, they pay more for less. In the mean time, underrepresentation in the RWUA and internal struggles driven by personal interests within some of the WUAs leave the members in a vulnerable position to secure access to water. Reflecting on the first narrative used to justify the reform process we observe that the increased water security did not lead to increased investment as the available amount of water is simply too little to push the agricultural production of the WUA members beyond subsistence level. Moreover, the policy model aimed at increasing the water security for private entities such as the WUAs, which did not translate into an increased water security for individual small-scale farmers within the irrigation systems. For the second narrative we can conclude that, even though small-scale farmers are now officially represented in the decision-making platforms, it did not lead to more effective decisions in terms of safeguarding their interests. The predetermined and biased organizational structure introduced by the policy model has left the WUA members without the necessary voice to participate meaningfully in decision-making processes. As a consequence of the fact that WUA members do not have modern infrastructure to limit loss of water, the payment for water did not lead to more efficient use of water within the irrigation system as was assumed within the third narrative. The small-scale farmers have simply too little control over the water to optimize their water use and maximize productivity. Moreover, the policy resulted in WUA members paying disproportionally more than they ought to pay on an individual basis. The narratives that were used to justify the reform process did thus not materialize for this group of water users; in fact the policy model has had a negative effect on access to water for most WUA members.

For those small-scale farmers who are not a member of a WUA their right to water abstractions from the river on an individual basis is virtually removed by the criteria set by the Likii RWUA. Even though it is not a statutory obligation to join a WUA, in reality this is what happens as articulated by one of the small-scale users: *"I do not submit an individual permit application because I am afraid that the RWUA will not recommend it and will force me to join the WUA"* (Interview F20, 2010). Due to the high costs paid for (access to) water by WUA members and the limited amount of water delivered through the WUA infrastructure it is not attractive for small-scale farmers to join the WUAs. As a result, some of them are forced to remain reliant on rain-fed agriculture for their livelihoods. The WUAs have set up some rules for WUA members to provide water to their unconnected neighbours mainly for domestic purposes. Even though this gives non-WUA members some security in (paid) access to water, it leaves them at the mercy of others: *"My neighbour sometimes declines giving me water when in a bad mood"* (Interview F32, 2010). Other more fortunate small-scale farmers with riparian access to land resort to pumping. These small-scale abstractions were tolerated under the previous Water Act when most small-scale water users informally fetched water. However, the 2002 Water Act labelled this kind of abstraction as illegal. This leads to recurring conflicts with the RWUA and WRMA who try to stop the abstractions during low river flow by confiscating the pumps. This notwithstanding and despite the relative high costs of pumps and fuel, it gives these small-scale irrigators the opportunity to access water based on crop needs, with some farmers irrigating plots of 0.8 hectare both during the dry and rainy season for commercial purposes. This does not only prove that the three narratives used to justify the reform process are invalid for this group of actors, but even destabilizes these narratives as some non-WUA members have increased their agricultural production compared to their neighbours despite decreased security in access to water, without voice in the participatory platforms and without paying water fees. It can be concluded that, even though the adopted policy model has excluded this group of small-scale farmers, it did not necessarily negatively affect access to water for some of them, while others feel inhibited because of the push to join the WUA.

5.8 Discussion

This paper shows that the narratives used to justify the reform process can only be upheld for some of the water users in Likii catchment and leaves the majority of the water users in the study catchment with a policy model that marginalizes them. These differential outcomes of the Kenyan water reform process can be partly explained through the interaction between the introduced public policies and the existing institutions at local level. The existing institutions are historically produced, intrinsically plural and unequal, creating different realities for the small-scale subsistence farmers and the foreign large-scale commercial farmers. The 'roll-back' of state services from provider to manager and the 'roll-out' of specific policy narratives and associated policy model are the products of a global policy network and disseminated to Kenya through the funding mechanisms of the World Bank and other agencies. Central elements of the policy model are securing property rights for private entities, decentralization of decision-making and economization of natural resources use, which are argued to be concurrently operating dimensions through which neoliberal shifts can materialize (Tickell and Peck, 2003; Harris, 2009; see also Bakker, 2007; Ahlers, 2010). Operating in an international business environment, these neoliberal inclined normative blueprints are more familiar to commercial farmers than to the small-scale farmers who thus far have faced a completely different institutional context. This does not only make it easier for the commercial farmers to adopt the new policies and adjust their practices but also to tweak the rules of the game in such a way that it better serves their interests (see also Kemerink et al., 2013).

91

Nevertheless, this does not explain the full story as the water reforms also have considerably different outcomes among the small-scale water users who operate within a similar institutional context. To understand this we need to not only look at the plural institutional landscape, but also the diversity in the physical environments in which the actors carry out their daily activities. The foreign-owned flower farms export their products to the European market and can reinvest their profits in innovative irrigation technologies to ensure effective and efficient water use. Moreover, their ability to invest in irrigation infrastructure makes it feasible for them to settle in the drier but less densely populated lowlands, allowing for larger farms with higher economic returns. The small-scale farmers in the midlands have less financial means to invest in hydraulic infrastructure. Whereas their necessity to collaborate was initiated by a physical imperative in terms of the collective action needed to construct and maintain the hand-dug furrows, with time it has shifted to an administrative imperative; first voluntarily in order to receive funding from NGOs to construct the piped network to distribute the water in the irrigation scheme and now under the water reform process reluctantly to maintain access to, and pay for, water use based on a collective permit. Even though a piped system might lessen the (collective) labour to maintain the system and reduce leakage compared with unlined furrows, the system is less flexible to adjust in case more water is needed. Moreover, open furrows have the advantage that it is easy to follow where the water flows and thus improve the monitoring and the transparency in water allocation among the farmers. The intake structures with control valves protected with three locks might have benefitted downstream users, but did little to protect the WUA members from misuse of water within their own network. The original design of the irrigation systems with a centralized storage facility to bridge 90-days dry spells seems difficult to achieve, since each WUA has a large command area which requires a large and thus expensive reservoir. A system of decentralized storage reservoirs seems a more practical option and offers more flexibility in terms of water distribution depending on local geographical conditions and cropping patterns (see also Van der Zaag and Gupta, 2008; McCartney and Smakhtin, 2010). Especially the decision of the WRMA to grant the WUAs water permits before any storage facilities are in place is questionable: while it has increased the revenue of WRMA, it has weakened the position of the members to demand further infrastructural investments from its leaders. The small-scale irrigators who opt not to be members of the WUA rely on the more flexible technology of pumping and can easily adjust the water intake based on the needs of their crops and the return on the investment. This allows them to move beyond subsistence farming and sell their harvest on local markets. Paradoxically, only through rejection of the neoliberal inclined water reforms and by opting not to be incorporated in the policy model, these small-scale irrigators manage to actively, and to some extent successfully, participate in the market economy. It should be noted that this is only feasible for farmers who have, for whatever reason, access to a composite set of resources including plots in close proximity to the river. Small-scale farmers who face less advantageous geographical conditions are either forced to join a WUA or, in case they cannot afford membership, are left to the mercy of the rain for their subsistence.

This paper shows that the water resource configurations in the Likii catchment are constituted by the interplay between a normative policy model introduced in a plural institutional context as well as the disparate infrastructural options and agricultural plots available to the various water users within the catchment (see also Swyngedouw, 2009). This interplay produces an uneven waterscape that is shaped by historically unequal, yet dynamic, social relations rather than following the simplistic and supposedly universally applicable causal relations assumed within the 'logics' articulated within policy narratives. In this process hydraulic infrastructure matters and therefore we argue that perhaps the most effective way of steering water resource configurations is revaluing, at least partly, the importance of physical control over water resources (see also Lankford, 2004; Swatuk, 2008; Van der Zaag and Bolding, 2009; Kemerink et al., 2011; Veldwisch et al., 2013). Policies are always politicized and based on (implicit) ideological preferences, whether they focus on infrastructure development or institutional change, and whether they serve the interests of the elites or protect the concerns of the marginalized. However, the shift in public policies towards solely steering institutional change processes has given bureaucrats responsible for water resource management less means to directly influence water resource configurations on the ground. Therefore it is in the interests of governments, especially those concerned with redressing the colonial past, to adopt a comprehensive approach in public policy that encompasses both physical as well as institutional components. After all, history has taught us that resource acquisition by European settlers in Africa did not only thrive as result of beneficial legislations but also as result of massive financial support from colonial authorities to develop hydraulic infrastructure to such extent that even decades after the colonial era "... *their rights that are fixed in permanent concrete structures such that the technology itself ... is able to do the work of social differentiation.*" (Mosse, 2008: 944; see also Manzungu and Machiridza, 2009). Direct investments in infrastructure for marginalized water users and targeting the actual (re)distribution of water to the users might be more effective for achieving equity than focusing exclusively on the establishment of 'inclusive' 'participatory' platforms, setting 'progressive' water tariffs or providing 'security' in access by granting conditional water use permits without effective monitoring of water use (see also Kemerink et al., 2013; Van Koppen and Schreiner, 2014). This also implies that we have to redefine the indicators that are selected to monitor the performance of a policy and include actual water flows and harvested crops. Measuring the number of members WUAs have, the number of women in executive positions, the number of meetings held, the number of administrative permits granted or the amount of water fees paid might be very informative for other purposes but has so far said little about the actual distribution of common pool resources among users. Moreover, actual investments in infrastructural development better justify the payment for water than the 'logic' put forward currently and might therefore increase the ability and willingness to pay. Would it only be a matter of time before a policy entrepreneur picks up these insights and spins infrastructure back into the global policy networks or is there really insufficient political will to redress inequity?

6. Jumping the water queue: changing waterscapes under water reform processes in rural Zimbabwe [46]

Abstract

Whoever visited Zimbabwe during the economic meltdown in 2008 would have been astonished by the daily neat lines of people patiently queuing day-in day-out for literally everything; from bread and eggs to cash and fuel. However, this unwearied behaviour of waiting for a turn was not observed during this research on the implications of the water reform process, initiated in 1998 and implemented during the economically unstable decade that followed. Instead, the case study presented in this paper shows that, those who could afford, jumped the queue by moving their agricultural activities upstream in a catchment to secure their access to water. This unforeseen response to the water reform process can be explained by adopting a socio-nature approach in which historically produced social relations and natural processes are conceptualized to simultaneously constitute and reorder physical environments forming dynamic yet uneven waterscapes. This paper discusses how satellite images can be used within the policy making process to capture the dynamic and context specific responses of actors to water reforms and as such give policy makers a tool to better monitor and steer the outcomes of policy interventions.

---

[46] This chapter is based on: J.S. Kemerink, N.L.T. Chinguno, S.D. Seyoum, R. Ahlers, P. van der Zaag (*forthcoming*) Jumping the water queue: changing waterscapes under water reform processes in rural Zimbabwe. Under review *Natural Resources Forum.*

6.1 Introduction

Whoever visited Zimbabwe during the economic meltdown in 2008 would have been astonished by the daily neat lines of people patiently queuing day-in day-out for literally everything; from bread and eggs to cash and fuel. However, this unwearied behaviour of waiting for a turn was not observed during this research on the implications of the water reform process, initiated in 1998 and implemented during the economically unstable decade that followed. Instead we observed that those who could afford jumped the queue by moving their agricultural activities upstream in a catchment to secure their access to water. This paper attempts to explain this unforeseen response to the water reform process by adopting a socio-nature approach in which historically produced social relations and natural processes are conceptualized to simultaneously constitute and reorder physical environments forming dynamic waterscapes. The question is how these complex processes can be incorporated in the policy making process to ensure policy makers can respond to dynamic and context specific circumstances?

This paper contributes to the ongoing discussion on the implementation of water reform processes in rural African waterscapes (Wester et al., 2003; Van Koppen and Jha, 2005; Kemerink et al., 2011; Mtisi, 2011; Kemerink et al., 2012; Manzungu, 2012; Van Koppen et al., 2014) by analyzing how the Zimbabwean water reform process triggered the reordering of a waterscape with disparate implications for the water users in the catchment. The Nyanyadzi catchment is used as case study for this paper and is located within the Save river basin in the eastern part of Zimbabwe. The findings presented are based on in-depth semi-structured interviews with 21 water users within the case study area and 6 government officials involved in the water reform process. The interviews were carried out between October 2011 and January 2012, with a follow-up visit to the catchment in 2013 and 2015. The interviewees were selected by a stratified random selection procedure to guarantee geographical spread and to obtain input from various categories of water users, as well as different age, class and gender groups. The findings of the interviews were cross-checked through focus group discussions, observations, analysis of relevant documents such as policies, meeting minutes and databases, comparison with existing literature and by consulting scientists and non-governmental organizations (NGOs) active in the region. In addition, the data collected through interviews on the physical changes within the waterscape have been compared with publically available satellite images.

This paper first explores the theoretical considerations by discussing the concept of socio-nature. In the next section a narrative of the catchment is provided including an analysis of the historical and institutional context. Thereafter the water reform process as envisioned at national level is described as well as how it unfolded within the case study catchment. This is followed by an analysis of how the reform process has physically changed the Nyanyadzi waterscape and what the implications are for the different groups of water users in the catchment. The concluding section reflects on the unintended and unforeseen outcomes of the reform process in Nyanyadzi catchment and discusses the implications for policy makers in charge of steering water reform processes.

6.2 Theoretical considerations

Building on political ecology, this article adopts the view that social relations and natural processes simultaneously constitute and reorder physical environments forming dynamic waterscapes (Swyngedouw, 1999; Castree and Braun, 2001; Budds, 2008; Mosse, 2008; Nightingale, 2011; Di Baldassarre et al., 2013). In this view, the waterscape can be understood as a historically produced socio-natural entity in which the environment is not regarded as simply *"a stage or arena in which struggles over resource access and control takes place ... [but where] nature, or biophysical processes, ... play an active role in shaping human-environmental dynamics"* (Zimmerer and Bassett, 2003:3). The waterscape is thus not only produced by human interventions in nature like the damming of rivers, the diversion of the water flows and the construction of infrastructures to distribute water to users (Loftus, 2007) and the responses of nature to this occupancy (Budds and Sultana, 2013), but also by the agency of the physical environment itself. After all, the domination of humans over nature is always incomplete and nature may 'strike back'. Thus ecological relations continue to shape and reshape societies and circumscribe the ever changing range of choices available for human exploitation. Organisms, not only human beings, have agency; they do not simply adapt to the environment they live in, but rather they are continuously constructing and destroying the world they inhabit and as such every organism affects its environment by causing it to change (Swyngedouw 2006; Moore, 2011). The agency of the physical environment is also exercised by (hydraulic) infrastructure. Not only does hydraulic infrastructure materially (re-)organize space, once constructed, it also becomes a force in itself, capable of rearranging and affecting water flows, often outliving the particular alliances who constructed it (Mosse, 2008). As such hydraulic infrastructure is not merely a passive instrument of human will, but an agent that actively opens certain trajectories while foreclosing other, alternative pathways in society (Swyngedouw, 1999; Ahlers et al., 2011; Meehan, 2014; Van der Zaag and Bolding, 2009). The waterscape is thus dialectically produced by actors of human and non-human nature in an ever ongoing process.

In the constant reordering of waterscapes, unequal social relations play a central role *"in determining how nature is transformed: who exploits resources, under which regimes and with what outcomes for both social fabrics and physical landscapes"* (Budds, 2008:60; see also Leach et al., 1999; O'Reilly et al., 2009). As such, waterscapes are never neutral but represent as well as shape political alliances and are continuously contested by rival coalitions (Haraway, 1991; Swyngedouw, 1997; 1999; Zimmerer and Bassett, 2003; Budds, 2008; O'Reilly et al., 2009; Budds and Sultana, 2013). In waterscapes uneven social relations materialize and consolidate, yet, as these inequities become visible in the landscape, the manifestation of these uneven relations can also be challenged, which ultimately might change these relations into new, not necessarily more even, social relations. Swyngedouw therefore argues that *"the flow of water ... embodies and expresses exactly how the 'production of nature' is both arena for and outcome of the tumultuous reordering of socio-nature"* (Swyngedouw, 1999:449).

Analyzing water reform processes from a social-nature perspective means recognizing that changes in water legislation and public policies do not only have implications for entitlements to water within a particular waterscape, but water users might also respond to these changing entitlements by reordering their physical environments. These changes in the physical waterscape often leave traces far beyond the political era in which the reform processes were initiated and do not simply disappear when a new piece of legislation is enacted or new policy objectives are formulated. Consequently the waterscape embodies*"...layer upon layer the legacies of former institutional arrangements, and of the changing environmental entitlements of socially differentiated actors"* (Leach et al., 1999:239). The outcomes of a water reform process are thus not only uncertain because they are implemented in an institutional plural context (Cleaver, 2002; 2012; Sehring, 2009; De Koning, 2011; Kemerink et al., 2013) but also because they are implemented in a intricate physical environment in which societal and natural responses to former political alliances have materialized. These complex processes make that water reforms seldom fully achieve the envisaged objectives and regularly lead to unintended consequences, both for water users as well as the environments they live in.

Swyngedouw argues that understanding waterscapes as historically produced outcomes of socio-nature processes, rather than based on particular rationales and principles articulated in public policies, calls for *'a transformation in the way in which water policies are thought about, formulated and implemented"* (Swyngedouw, 2009:56). However, in their very essence policies are always based on simplified models of reality and incorporating complexities within policy making process has proven very difficult to achieve (Cleaver, 2002; Mosse, 2004; Rap, 2006; Molle, 2008 Laube, 2010; Peck and Theodore, 2010; Bourblanc, 2012; Kemerink et al., *forthcoming a*). In addition, policies are not implemented in isolation; often they coincide with reforms processes in other domains or they are affected by economic or political developments in society.

In this article we argue that analyzing the interactions between physical and social changes in waterscapes can be an important approach to enrich policy studies and can aid policy makers who are faced with the discrepancies between what is written on paper and what happens on the ground. It can reveal to what extent envisioned reforms materialize within waterscapes and how people respond to these changes by reordering and transforming their physical environments. Moreover, it can explain uneven outcomes of water reform processes within a particular waterscape. It may also reveal how geographical conditions in catchments shape water reform processes as well as capture how nature responds to the reordering of waterscapes as a result of the reforms. In this way, it can establish the environmental consequences of water reforms, for instance alteration of river flows and groundwater levels, changes in the frequencies of floods and droughts or changes in river morphology. We thus argue that studying physical changes in the waterscape from a socio-nature perspective might give policy makers a holistic insight in the implications of their interventions and allow them to respond to local dynamics, or even readjust their policies altogether, in case of unintended detrimental consequences. In this article we will use the concept of socio-nature to analyze the historical production of the Nyanyadzi waterscape and specifically to unravel the implications of, and responses to, the 1998 water reform process.

6.3 Setting the scene

The Nyanyadzi river is the eastern part of Zimbabwe is a tributary of the Odzi River just before the confluence of the Save River (Figure 6.1). The water within the Nyanyadzi catchment originates from the Chimanimani mountains on the border with Mozambique. The Nyanyadzi catchment carves out an 800 km^2 area with an average rainfall of 1,200 mm/year in the upstream part of the catchment and less than 500 mm/year downstream (Magadlela, 1999; Bolding, 2004). Most of the streams within the catchment are perennial with extreme low flows during winter, from May to August, while most rain falls in summer between November and March. The upstream part of the Nyanyadzi catchment generally has rich loamy soils and with the abundant rainfall is ideal for intensive maize cultivation and fruit production. However, the middle and downstream parts of the catchment generally have poor sandy soils and farmers use fertilizers and rely on supplementary irrigation to realise a harvest.

Figure 6.1: location of Nyanyadzi catchment (ZINWA, 2012)

Contrary to claims of the first European settlers at the time that they arrived in the early 1890s, the upstream parts of the Nyanyadzi catchment were densely populated by people of the Ndau tribe while in the lower parts of the catchment Ndau settlements were clustered along the river and perennial tributaries (Bolding, 2004). The Ndau mainly relied on a combination of rotational agriculture of the dry lands and permanent cultivation of the wetlands along the river banks. In some locations hand-dug furrows were used to divert water from the rivers for supplementary irrigation (Bolding, 1996). The agricultural land in the catchment was customarily owned by the chief and plots were allocated to families for farming. No customary rights were specified for ownership of grazing land, forest and water resources (Chikozho and Latham, 2005). Water was regarded as a divine gift and annually rain making ceremonies took place to worship ancestors (Magadlela, 1999; Vijfhuizen, 1999).

Upon their arrival, white settlers lodged land claims with the British South Africa Company. They carved out large farms of 2,500 hectares in the most fertile upstream parts of the catchment where they initially cultivated maize and grazed cattle for subsistence. Small pockets of less fertile lands were declared native reserves on which the Ndau were allowed to settle (see also Zawe, 2006). Within the reserves land was held under a tenure regime that was perceived by the settlers as 'traditionally African' in which the indigenous population only enjoyed user rights, while the settlers obtained full ownership rights to the land they were allocated (Jaspers, 2001; Zawe, 2006; Manzungu and Machiridza, 2009). In the native reserves rain-fed agriculture was practised; growing drought resistant small grains like millets and sorghum as well as groundnuts. In addition, small livestock like goats were kept for meat and milk consumption. Dry spells caused regular crop failure and only during good rainy seasons was the harvest sufficient for subsistence, forcing many men to work as labourers in the mines and on white-owned farms. Ndau families who remained living on their ancestral lands outside the reserves became rent-paying tenants by paying money or by providing three months of labour on the white-owned farms per year (Bolding, 2004). Most of these tenants continued their traditional practice of combining rain-fed agriculture with cultivating crops on small beds in the wetlands along the tributaries of Nyanyadzi River. Families cultivated different kinds of vegetables on two to three beds in gardens of about 0.05 hectares (Bolding, 2004).

In 1923 Zimbabwe became a self-governing colony of the British Empire and was formally referred to as Southern Rhodesia. The colonial authority endorsed a Water Act in 1927 that shifted the priority of water use for mining, as stated in earlier water ordinances, to irrigation. In addition, water for primary use was defined to secure access to water for basic human needs, which included gardening. In the 1947 Water Amendment Act this primary use was quantified as 228 litres per person per day. The Water Act of 1927 differentiated between public and private water with private ownership rights granted based on three main principles (Jaspers, 2001; Manzungu and Machiridza, 2009; Mtisi and Nicols, 2003; Mtisi, 2011):

- Riparian right principle: only owners of land could apply for rights to any water body on or directly adjacent to their land. This entailed that water rights were attached to land and not to individuals, which basically meant that water rights were automatically transferred to new land owners in case land was sold.
- Priority date principle (also known as the prior appropriation principle): ownership rights to water were granted based on first-come first-served basis with earlier granted rights given priority over new applications.
- Perpetuity principle: ownership rights to water were granted for an infinite time and could only be revoked under special circumstances such as allocating water to (nationally) strategic uses during droughts or when right-holders renounced their right.

In 1934 the colonial government started with the construction of the Nyanyadzi Irrigation Scheme in the most downstream part of the catchment. Although originally established to mitigate recurrent famine and developed with the aspiration to demonstrate the potential of 'modern' African agriculture, this scheme facilitated the removal of the native African population from the fertile upstream lands without making them dependent on the colonial state for food security (see Bolding, 2004, for a detailed narrative). As such the scheme, like similar schemes established elsewhere in the country, formed part of the implementation of legalized racial segregation that was introduced by the colonial authority since the mid 1920s in favour of the minority white population (Manzungu and Machiridza, 2009). Nyanyadzi scheme started off with one block, which is currently referred to as block C, but expanded into three more blocks further downstream (Block A, B and D, Figure 6.2). Once the scheme was completed the command area of the irrigation scheme covered 412 hectares in total with the first generation of farmers irrigating 1.0 to 1.2 hectares each (Bolding, 2004). Similar to the tenure regime in the reserves, the families that were settled in Nyanyadzi scheme obtained user rights, not full ownership rights, over the land they got allocated. It was envisaged that, once the scheme would be productive, the scheme would be controlled and maintained by the smallholder farmers on a communal basis. Nevertheless, when they refused to repair the infrastructure, lease fees for land including water to irrigate, and later levies on harvest, were introduced, turning the famers into plot holders within a government operated scheme.

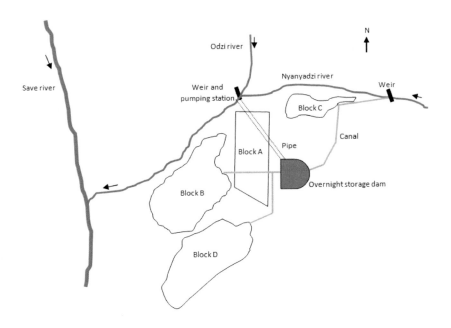

Figure 6.2: sketch of the Nyanyadzi Irrigation Scheme

In 1937 the irrigation scheme obtained the first water right within the Nyanyadzi catchment under the 1927 Water Act allowing the scheme to abstract 283 litres per second to irrigate 400 hectares. Even though the then prevailing water act recognized the priority date principle, the priority given to the irrigation scheme was limited to the drainage area covered by the upstream native reserves to ensure that white farmers further upstream in the catchment could still apply for water rights without considering the water right of the Nyanyadzi irrigation scheme (Bolding, 2004). The river water was diverted at a weir that was built across the Nyanyadzi River seven kilometres upstream of Block C and was conveyed by way of a gravity system of canals and gates to the fields. With time the water intake decreased, as a result of the increased (unofficial) water use by the upstream located European settlers, while the command area expanded and thus water availability became a serious concern within the irrigation scheme. Since 1942 there have been plans to put a dam on Nyanyadzi River to alleviate the water scarcity within the scheme (Bolding, 2004). Finally, in 1957 blocks A, B and D obtained additional water, released from Osborne Dam 200 kilometres upstream on the Odzi river, to address the most pressing needs. This water was initially pumped by diesel-run engines from a weir at the confluence of the Odzi and Nyanyadzi rivers which was conveyed by a pipe to Block A. Later the pumping system was changed to run on electricity and the pipe was extended to divert the water to an overnight storage dam from which it was channelled to blocks A, B and D (Figure 6.2).

Within the Nyanyadzi irrigation scheme mainly maize was grown in summer, while wheat and beans were cultivated during the dry winter months. In addition, profitable vegetables and fruit trees were planted on the edges of the plots along the irrigation canals. In the first decade the agricultural production of the irrigation scheme was limited, but from the 1950s the yields increased. Meanwhile, water rights were granted between 1939 and 1952 to ten farms located upstream of Nyanyadzi, allowing the farmers to abstract 77 litres of water per second to irrigate 114 hectares (Bolding, 2004). Even though the legislation favoured the white minority in the appropriation of water and land resources, it was not until other measures were put in place in the 1940s that a white-owned agricultural based economy started to flourish. These measures included subsidies and soft loans for the construction of hydraulic infrastructure, funding for soil conservation measures and secure prices for agricultural products up to 40% above the market value (Manzungu and Machiridza, 2009). In the mean time plot sizes in the Nyanyadzi irrigation scheme decreased as more Africans were settled into the scheme and plots were divided among younger generations, forcing the irrigators to cultivate additional land outside the irrigation scheme or search for alternative livelihoods.

Resistance against the colonial state got momentum during the 1960s and 70s and, due to its close proximity to Mozambique which obtained independence in 1975, the catchment became an important nursing ground for nationalist political movements such as the Zimbabwean African National Union (ZANU)[47]. Redressing the racial dispossession of land became the main slogan to mobilize people into the armed struggle against the colonial regime (Zawe, 2006). Already before independence the district in which Nyanyadzi catchment is located "... was more or less run by the district ZANU party committee, that permitted groups of interested African smallholders to settle on the vacated European farms ... By the mid-1990s very few commercial farms were left, a distinctive feature of the district when compared with other districts" (Bolding, 2004: 28). A few years after independence in 1980 the ZANU led government took over some of the remaining white-owned farms in the middle reaches of the Nyanyadzi catchment and created a smallholder resettlement scheme near the village of Zimunda (Figure 6.3). In this settlement scheme farmers divert water from the Nyanyadzi River via unlined canals locally referred to as furrows. The establishment of this scheme was corroborated in an interview with a resident from Zimunda village who said "soon after independence in 1980 we were told to register our names for consideration for land redistribution. I was one of the first people to be settled here in 1983, with the government coming in with tractors and earth-moving equipment to clear and flatten the land to create fields for us" (interview HD3). The settled families were given full ownership of the land through private title deeds, however only few plots came with water rights that were obtained by the previous owners of the land. To secure access to water thirty furrows in the Nyanyadzi catchment were granted water rights in the 1980s and early 1990s, including nine furrows in the Zimunda scheme. However, in practice, an estimated hundred furrows in the catchment extracted water from the river, irrigating up to 250 hectares of land (Bolding, 2004). Often water in the furrows was shared with other farmers, either based on kinship relations or conditional to in-kind contributions such as parts of the harvest or providing labour to clean the furrow. The Zimunda farmers agreed on a water sharing arrangement in which, during low flows in the river, the furrows abstracted water on a rotational basis (see also Kemerink et al., 2009).

Even though attempts were made to increase the involvement of farmers in the operation and maintenance, the Nyanyadzi irrigation scheme had for various reasons been controlled by the

---

[47] In 1988 ZANU become the Zimbabwe African National Union – Patriotic Front (ZANU–PF) after a merger with the Zimbabwe African People's Union (ZAPU). It has been the ruling party since independence led by Robert Mugabe. Only after the 2008 elections, in which ZANU-PF lost its majority in parliament, Zimbabwe was governed by a government of 'national unity' including ministers from ZANU-PF as well as the main opposition party. However ZANU-PF remained in control of key ministries including agriculture. In 2013 ZANU-PF obtained two-third of the seats in parliament again and re-established its authority. The results of the various general elections for parliament and presidency have been highly contested with reports on election fraud and violence against constituents (Sithole and Makumbe, 1997; Kriger, 2005).

government since its establishment. However, in 1987 the fiscal deficit forced the government to handover most of the management responsibilities under the so-called irrigation management transfer policy (Bolding, 2004; see also Rap, 2006; Zawe, 2006; Kadirbeyoglu and Kurtic, 2013). For the handover of the responsibilities from government to the farmers it was mandatory for the farmers to elect an Irrigation Management Committees (IMC). The IMC of the Nyanyadzi irrigation scheme comprised of eight elected members from the farmers community, two representatives from each block. Elections were held every five years, which were overseen by extension officers from the government. Even though the farmers had paid water fees for several decades already, now they became fully responsible to cover the costs for maintenance and operation of the scheme, including the payment of the electricity bill of the pumping station that supplied water to blocks A, B and D. The recurrent water shortages in the scheme together with the increased costs for accessing water led to tensions between the farmers in the Nyanyadzi irrigation scheme and the upstream water users, especially the Zimunda furrow irrigators. Between 1984 and 1994 several raids took place in which the Nyanyadzi irrigation scheme farmers destroyed the upstream structures, not only attacking illegal furrows but also furrows that had official water rights, claiming that the Nyanyadzi irrigation scheme had priority rights over the water in the river based on the principles embedded in the colonial legislation (Bolding, 2004).

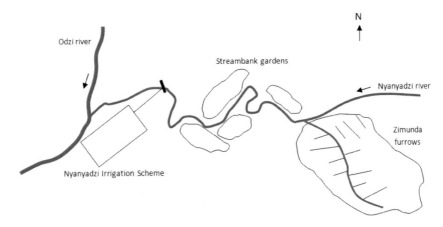

Figure 6.3: sketch of the location of the main water user groups in Nyanyadzi catchment since independence (1980)

6.4 The Zimbabwean water reforms

In 1998 new water legislations was enacted with the view to reform the water sector and specifically to redress the inequities in access to water (Jaspers, 2001; Manzungu and Kujinga, 2002; Swatuk, 2005). The new legislation abolished private ownership of water and introduced water user rights that are acquired through conditional permits (Manzungu and Kujinga, 2002; Manzungu, 2004). Even though existing lawful use of water was recognized under the new Water Act, the prior appropriation right as well as the riparian right were abolished (Jaspers, 2001; Makurira and Mugumo, 2003). For any water use, other than primary water use, a permit was required. Primary use is defined in the Water Act as *"reasonable use of water for basic domestic needs in or about the area of residential premises, animal life (other than fish in fish farms or animals or poultry in feedlots), for making bricks for the owner, lessee or occupier of the land concerned or for dip tanks"* (GOZ, 1998a: 3). Under the Water Act permit holders have to pay for their water use. The Water Act stipulates that the tariff, and changes therein, have to be justified based on *"the cost of providing, operating or maintaining the service concerned, any proposed improvements to any service or facility; and any other relevant economic factors"* (GOZ, 1998b:30-2). The Water Act also provides room for progressive tariff setting by stipulating that *"... different charges may be fixed for the sale of water to different classes of persons or for different uses. Provided that, in fixing different charges in respect of different classes of persons, there shall be no discrimination between persons on the grounds of race, tribe, place of origin, political opinion, colour, creed or gender."* (GOZ, 1998b:30-5). Moreover, under the new legislation the environment is considered as a rightful water user with an allocation reserved in river systems to maintain ecological integrity and adoption of the polluter pays principle.

Under the 1998 Water Act seven river catchments were identified based on hydrological boundaries, each to be governed by catchment councils. The council members are elected and/or appointed stakeholder representatives tasked amongst others with developing integrated plans for the catchments, revising and reviewing water allocation, issuing permits and collecting water levies (GOZ, 1998a:21; see also Manzungu and Kujinga, 2002). The councils have the authority to revoke or revise permits as they see fit, for instance in case of over-abstraction of the available water resources or in case of inequity in allocation (GOZ, 1998a:28). Each river catchment is divided in to sub-catchments in which sub-catchment councils have the task of regulating and supervising the lawful use of water, being responsible for the day-to-day management of the water resources in the catchment (GOZ, 1998a:24; see also Jaspers, 2001). The sub-catchment councils are the prime body through which permit holders pay fees and sub-catchment councils are entitled to levy additional fees for services they provide. Members of the sub-catchment council are elected and should include representatives of the different stakeholders in the area under its control. At national level the Zimbabwe National Water Authority (ZINWA) has been established with as key role to advise the minister responsible for water on the formulation of national policies and frameworks relevant for the planning, management and development of the country's water resources. ZINWA also manages the Water Levy Fund that is meant to promote water resources development and water service provision, in particular for the construction of hydraulic infrastructure in areas that have been disadvantaged during the colonial era (Makurira and Mugumo, 2003). The fund's revenue comes from the fees paid by permit holders and any other money the government receives or allocates for managing water resources (GOZ, 1998b: 33-39).

The implementation of the water reform process in Zimbabwe coincided with changes in the land reform policies in an attempt to speed up the redistribution of land from the white settlers to the native African population as little progress had been made since independence. Under pressure of international donors, the land reform process had so far been bounded by neoliberal principles such as willing-buyer, willing-seller and compensation of loss of property and income according to market prices. In 1992, the government adopted an act that abandoned the willing-buyer, willing-seller principle to force the acquisition of white-owned property to resettle smallholder farmers. However, when progress remained slow, frustration led to illegal, though government-encouraged, land invasions of white-owned commercial farms. The land appropriated under this so-called fast-track land reform programme was nationalized in 2005, depriving the former landowners of the right to appeal in court against the expropriation of their land or demand financial compensation (Zawe, 2006; Svubvure et al., 2011). In response to this fast track programme, international donors, who provided financial and technical aid to implement the water reform processes, withdrew their support from Zimbabwe. The sudden absence of donor support, as well as the economic meltdown that followed, stalled the implementation of the water reform process (Makurira and Mugumo, 2003).

6.5 Unfolding the water reforms in Nyanyadzi catchment

As result of the water reforms since 1999 the Odzi Sub-Catchment Council (OSCC) became the main regulator of water in Nyanyadzi catchment (Kujinga, 2002). The OSCC has an office in Mutare, a town 100 kilometres away from Nyanyadzi catchment, and falls under the authority of the Save Catchment Council (Kujinga and Manzungu, 2004). Even though the Water Act specifies that the members of the sub-catchment council should be elected representatives of stakeholder groups (GOZ, 1998a:24), many smallholder farmers in Nyanyadzi catchment do not know who represents them at the OSCC. According to the records of the OSCC, the farmers in the Nyanyadzi scheme are represented through the structures of the IMCs of the various smallholder irrigations schemes in the Odzi catchment, while the Zimunda resettlement farmers are represented through the Zimbabwe Farmers Union (Kujinga, 2002). The small-scale farmers who are abstracting water without permit are not represented at the OSCC. The OSCC member who currently represents the farmers in the Nyanyadzi irrigation scheme is member of the IMC of another irrigation scheme and was appointed without consultation of the Nyanyadzi IMC after the previous councillor passed away (interviews AO1, OSCC3). So far she has not visited the Nyanyadzi irrigation scheme and is unknown by the farmers. The OSCC claims that some of the councillors long lost their legitimacy as they fail to represent their constituencies by not communicating to them what has been discussed and agreed in the council meetings and vice versa, while the councillors complain that they do not have enough resources to go around the catchment to hold consultations with the farmers they represent (Chinguno, 2012; Kujinga, 2002).

Moreover, from analyzing the minutes of the Save Catchment Council and the minutes of the OSCC as well as attending the 2011 OSCC Annual General Meeting it becomes clear that the council so far hardly discusses content issues related to water, instead the meetings mainly focus on managerial issues. A few councillors, mainly those representing the white commercial farmers in the Odzi catchment, are dominating the meetings and continuously raise questions about administrative and financial matters (see Chinguno, 2012, for a detailed narrative; see also Kujinga, 2002; Kujinga and Manzungu, 2004). As a result catchment plans for water allocation have not yet been developed nor have monitoring plans for water use been implemented. Another drawback for developing catchment plans is the outdated database in which water permits in the catchment are registered; for circa 13% of the total 4,162 permits issued in the Save catchment the current permit holder is unknown as well as the amount of water they are entitled to. This applies to four out of the forty-one water permits issued within the Nyanyadzi catchment. This ambiguity in permit holders is largely the result of rapid and somewhat obscure changes in the title deeds under the fast-track of the land reforms. For the other permits the permit holders are known, though it is unknown if the actual abstraction is according to the permitted use and if the permit holders have fulfilled their duties such as paying the applicable fees. The bias towards internal organizational matters rather than implementing the council's core activities of regulating water use and supervising the day-to-day management of the water resources also becomes clear from analyzing the OSCC budget allocation for 2012. The expected income for 2012 that would be collected in the Odzi catchment from water fees paid by permit holders was estimated at about

300,000 US$ (excluding an estimated 265,000 US$ of arrears from non paid fees). As the OSCC has to pay approximately 100,000 US$ to the national Water Levy Fund, it is left with an annual budget of about 200,000 US$. The 2012 budget shows that about 47% was expected to be spent on staff salaries, meeting venues and allowances for OSCC councillors, 18% was allocated for office consumables, office equipment, administration and contingencies, 24% for transportation, including purchasing of council vehicles, and the remaining 11% for awareness workshops and the annual general meeting.

How the money of the Water Levy Fund is being spent at national level remains unclear. The ZINWA website mentions that two dams were under construction, the construction of one started in 1998 but is yet to be completed while the water from the other dam is not yet been used since its completion in 2004 as secondary infrastructure to distribute the water to the users is absent. Nevertheless, despite these impediments, on the ZINWA website government officials make pledges for the financial support of the construction of more dams. It should be noted that the official purpose of these dams is only partially to distribute water to smallholder farmers as the dams will also supply water to sugar cane estates and/or towns and/or used for hydropower generation. In one case, a dam that will be constructed with support from a Chinese company, the beneficiaries are not explicitly stated. None of these dams are located in the Save River basin and by 2015 no information was publically available on any government investment in water resources development within the basin. In response to the OSCC budget allocations the Save Catchment Council Manager stated that *"councillors and catchment council staff must value the water users as they are the life givers of the catchment and sub-catchment councils"* (intervie OSCC, 2010). The farmers, however, do not understand why they need to pay water fees while they do not receive any services from the OSCC that improve their access to water, which is according to the Water Act a prerequisite for charging water fees. A traditional leader argued in a meeting with the water authorities on the topic of charging water fees: *"neither the water authority ZINWA nor the relevant sub-catchment, OSCC, has built a dam nor added anything to the water"* (interview CH1). Similarly a farmer stated in an interview that *"their (ZINWA and OSCC) message is always pay, without explaining what the money is for. They forget that the river is ours. Had it been they had put a dam on the Nyanyadzi River, we would understand what the money is for"* (interview FGD7). In addition to the disagreement on budget allocations, several incidences on financial mismanagement within the OSCC had been reported during the preceding years including the payment of bonuses to councillors without prior approval and the omission to account for the sale of a vehicle owned by the council (see Chinguno, 2012, for a detailed narrative).

The collaboration over water among the actors in the catchment is further complicated by political rivalry between the ruling party ZANU-PF and the main opposition party, especially after the 2008 elections in which ZANU-PF lost its parliament seat for this constituency (interviews PL1, PL2, IFA1, IFB2, NGO2, FGD6, AO1). In particular the IMC has been accused to align with the opposition party, which has sparked unwillingness of farmers to be associated with, and attend meetings of, the IMC out of fear for reprisal (interviews PL1, IFA1, HD4) and ZANU-PF activists threatening IMC officers carrying out their daily duties (interview AO1).

In the mean time, the physical access to water has changed for the water users within the catchment as result of the water reforms. The least has changed for the furrow irrigators around Zimunda. The existing water rights that nine furrow irrigators obtained under the previous legislation have been automatically converted into water permits. Even though the permits are conditional on payment of the water fees, the irrigators so far have not paid for the water they abstract as they claim that most of the harvest is used for subsistence rather than commercial purposes (interviews HD1, FG2). With no metering system in place to monitor the actual abstractions by the furrows, it is difficult for the authorities to enforce the payment for water. Another eleven Zimundo irrigators, who did not have water rights under the previous water act, have continued to abstract water without formal entitlement (interviews HD3, FGD2). Even though they risk getting a penalty for abstracting water without permit, the limited monitoring of the water abstractions by the OSCC have so far allowed them to continue their water use practices. The water used by farmers who cultivate gardens on the banks of the rivers used to fall under primary use which was exempt from requiring a water right, however, their water use is deemed illegal under the 1998 Water Act since it is not included in the definition of primary use (GOZ, 1998a: 3). Moreover, the Environment Management Agency claims that the practice of irrigating on river banks causes erosion, soil degradation and siltation of the river bed. As result these farmers are under pressure to give up their farming activities and in response they have resorted to abstracting water at night to avoid confrontation with the authorities (interviews G1, AO1, G2). The water reforms have had the biggest implications for the farmers in the Nyanyadzi irrigation scheme. They enjoyed, albeit limited, priority rights under the previous Water Act, a privilege that under the 1998 Water Act has been taken away from them. Being located most downstream in the catchment, they have moved from first position to the last in the water queue on the Nyanyadzi River. This has especially affected the water availability in Block C that fully depends on the river for its supply. Furthermore, since the irrigation scheme has a gauging station at the river intake and the pumping station is also equipped with a water meter, it is easier for the authorities to charge the irrigation scheme fees for the abstracted water compared to the unmetered users upstream. Moreover, rather than dealing with a large number of individual farmers, the authorities can put pressure on a single entity, the IMC, to pay the fees, while the IMC in their turn can refuse to allocate water to farmers who have not fulfilled their obligations[48]. This has forced farmers to actually pay the water fees or to bribe the water operators, making the price they pay to access water relatively high compared to other water users in the catchment. The price Nyanyadzi irrigators pay continues to increase; from 5 US$ for irrigation one acre of land in 2012 to 17 US$ in 2015.

The situation has especially worsened for the farmers in the irrigation scheme since the dollarization of the Zimbabwean economy in 2009 in an attempt to combat the hyperinflation as result of the economic meltdown. By the end of 2011 the Nyanyadzi irrigation scheme farmers had built up a collective dept of US$ 28,000 with the electricity company. As a result the electricity supply was disconnected in October 2011, leaving the irrigation scheme solely relying on the little water that was still flowing in the Nyanyadzi River. Since then only Block C received some water to irrigate plots, while the other blocks completely dried up. The acute water shortage also resulted in an outright refusal to pay water fees, accumulating the debt of the irrigation scheme with ZINWA to US$ 17,000 for unpaid water fees. Even though both debts have been cancelled during the campaigns for the 2013 presidential elections in an

---

[48] ZANU-PF affiliated farmers argue that the IMC is implementing the political agenda of the opposition party, including charging high fees for water use and harassing farmers to pay, and claim that the ZANU-PF led government does not demand historically marginalized farmers to pay and instead provides handouts to them (interviews AO1, PL1). Whereas farmers affiliated with the opposition argue that the excessive amount of money they need to pay to access water is a punishment by ZANU-PF because the farmers voted for a candidate from the opposition party during the 2008 general elections (interviews HD4, IFA1).

attempt to win back the constituent seat, the situation did not improve due to a breakdown of the pumps while there was insufficient funding for repairs.

6.6 Reordering the Nyanyadzi waterscape

The changes in institutional arrangements during the last century have changed the waterscape of Nyanyadzi catchment. In the pre-colonial time mainly the upstream part was inhabited and the natural available soil moisture in the river banks was permanently used for gardening, while the soil moisture in the drier parts of the catchment was only used on a rotational basis for subsistence farming. Only a few unlined furrows were constructed for supplementary irrigation by the smallholder farmers. The colonial time was marked by a move downstream as result of forced relocation of the indigenous population. A racially segregated waterscape was produced through concrete hydraulic infrastructures that provided water to the white-owned large-scale commercial farms in the upstream parts of the catchment and the communal smallholder plots in the downstream irrigation scheme. In between the irrigated plots, soil moisture became permanently used for subsistence farming both in the drier parts of the catchment as well as on river banks. The post colonial era manifested in the waterscape through the division of the large-scale farms into smallholder plots and the increase of lined and unlined furrows in the upper and middle reaches of the catchment for irrigation throughout the year. This resulted in water scarcity within the downstream irrigation scheme, especially for the farmers located in Block C during the dry winter months.

The water reform process initiated by the 1998 Water Act has also left its marks on the Nyanyadzi waterscape. The changes in entitlements and the difference in costs for accessing water as result of the reform process in combination with the absence of investments in hydraulic infrastructure, has triggered a move upstream; farmers who managed to obtain land through the traditional tenure system have jumped the water queue by leaving the downstream irrigation scheme and establishing irrigated plots further upstream along the river banks. These farmers often rent out the life-long lease they have on the plots in the irrigation scheme, while in some cases the user rights on the land have even been 'sold' to politically well-connected farmers for indefinite periods[49] (interview IAF4). The farmers who could not afford to jump the water queue have started to practice rainfed farming within the irrigation scheme. A farmer explains: *"we now practice dryland agriculture in our irrigation scheme and the scheme has since relocated to the river"* (interview G1). The proliferation of irrigated plots along the river banks and the drying up of the downstream irrigation scheme is visible on satellite images. Figure 6.4 shows the Normalized Difference Vegetable Index (NDVI) on processed satellite images of the river section directly upstream of the weir that diverts the water into the Nyanyadzi irrigation scheme. The NDVI value ranges between -1 and 1 with values above 0.4 indicating green (living) vegetation and values below 0.4 indicating bare soil or very dry (non-living) vegetation. The satellite images have been made mid September which is towards the end of the dry (winter) season and as such all green vegetation must be either irrigated crops or riparian vegetation. Comparing the images of 2005 (Figure 6.4a) with

---

[49] Officially the land under lease cannot be sold or rented out to third party; the lease-deal can only be renounced and returned to the District Administrator or the District Administrator can revoke the lease if the plot holder fails to work the land. The District Administrator has so far tolerated the private obscure transactions that are inconsistent with the law, and thus somehow legitimized the transfer of land to political well-connected individuals, despite the existence of a waiting list for families who would like to be allocated land within the irrigation scheme (interview IAF4).

the image of 2014 (Figure 6.4b) shows a considerable increase of green vegetation along the riverbed during the last decade. Arial pictures confirm that this are irrigated gardens rather than an increase in natural riparian vegetation (Figure 6.4c).

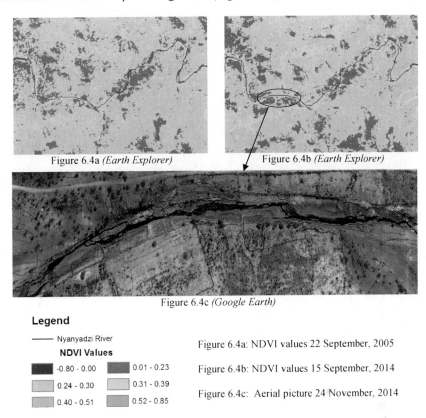

Figure 6.4a *(Earth Explorer)*   Figure 6.4b *(Earth Explorer)*

Figure 6.4c *(Google Earth)*

Legend

—— Nyanyadzi River

NDVI Values

-0.80 - 0.00	0.01 - 0.23
0.24 - 0.30	0.31 - 0.39
0.40 - 0.51	0.52 - 0.85

Figure 6.4a: NDVI values 22 September, 2005

Figure 6.4b: NDVI values 15 September, 2014

Figure 6.4c: Aerial picture 24 November, 2014

Figure 6.4: Satellite images of Nyanyadzi River section upstream of intake of Nyanyadzi irrigation scheme

Whereas in the irrigation scheme both the winter and the summer harvests regularly fail due to a shortage of water, the plots along the rivers can harvest up to three times per year. Not only has the water distribution in the waterscape been altered as result of the reform process, also the plot sizes have changed. Whereas in the past the average plot sizes in the Nyanyadzi irrigation scheme decreased as result of subdivision of plots amongst heirs (Bolding, 2004), the average plot sizes have increased since the water reforms as a few people have 'bought up' the land from people who left the irrigation scheme. These actors who obtain the vacant land are politically influential persons affiliated with ZANU-PF who, through patronage, manage to obtain the little water that is still available in the irrigation scheme and/or who anticipate the availability of water within the irrigation scheme in the near future (interview IFA4). Along with these new actors, traders entered the irrigation scheme and now control considerable parts of the agricultural business in and around the scheme, including changing the main cropping pattern to sugar beans and tomatoes. In the mean time, the plot sizes along the river have also considerably increased, where it used to be small vegetable gardens now demarcations of individual plots up to 0.5 hectares are visible (Figure 6.5). Some of these plots are not located on the banks of the river, but in the river bed itself (Figure 6.5a and 6.5c).

This has potentially detrimental consequences for river flows and soils, but also leaves these farmers vulnerable to destruction of their crops by seasonal flash floods as occurred for instance in December 2011. It should be noted that these plots along the river are not exclusively in use by farmers who moved out of the Nyanyadzi irrigation scheme, but include people coming from (peri-)urban settlements in search for alternative livelihoods in response to political turmoil and economic decline (Bratton and Masunungure, 2006).

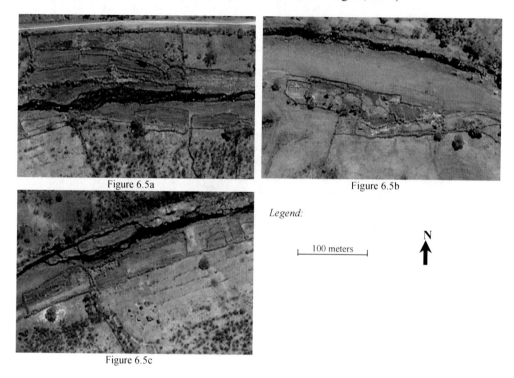

Figure 6.5a   Figure 6.5b

Legend:

|___ 100 meters ___|

N

Figure 6.5c

Figure 6.5: Aerial pictures of irrigated gardens upstream of the intake of Nyanyadzi irrigation scheme (Google Earth, 2015)

The financial challenges within the Nyanyadzi irrigation scheme also reordered the waterscape. Whereas in the past Block C was disadvantaged as it only received water from the dwindling Nyanyadzi River, now it is the other blocks in the irrigation scheme that struggle more; no water is pumped anymore from the Odzi River and the water from Nyanyadzi River is fully used by the farmers in Block C. This is clearly visible on the aerial pictures in Figure 6.6, where during the start of the winter cropping season of 2013 Block C shows several irrigated plots on which crops grow (Figure 6.6a), while Block A directly downstream does not show any sign of irrigated crops and has completely ran dry (Figure 6.6b). Field data confirmed that also during the 2011-2012 summer cropping season Block C was the only block to realize a reasonable harvest, while in the other blocks the harvest almost completely failed as result of water scarcity. Farmers in Block C argue that they do not let water flow to the downstream blocks as the little water available in the system will not reach these blocks (interviews IFD1, IFD2, IFB1, IFA3), though some interviewees also reveal that farmers in Block C claim priority rights over the water originating from Nyanyadzi river. Block C farmers claim that they are the decedents of the first group of farmers that settled in the irrigation scheme and, since their forefathers vacated ancestral land to make place for the scheme and the irrigation canal that diverts water from Nyanyadzi River, they reason that they have priority rights to the water (interviews FGD8, FGD1, HD2, HD4; see also Bolding 2004).

The NDVI values towards the end of the cropping season in September 2013 confirm that hardly any irrigation took place that year as only few signs of green vegetation are visible (Figure 6.6c). The cancellation of the debts for electricity bill and water fee in the run-up for the presidential elections in July 2013 came too late to avoid failure of the 2013 harvest. The NDVI values in the same month a year later, 2014, show a completely different picture; irrigated plots are clearly visible even in the tail-end of the irrigation scheme (Figure 6.6d). It thus seems that the government intervention to revoke the arrears benefitted especially those actors who obtained the land vacated by farmers that left the irrigation scheme in the preceding years in response to the multiple failures of the harvest due to the acute water shortage.

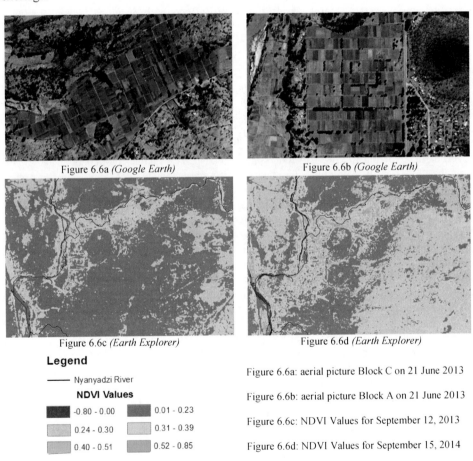

Figure 6.6a *(Google Earth)*   Figure 6.6b *(Google Earth)*

Figure 6.6c *(Earth Explorer)*   Figure 6.6d *(Earth Explorer)*

Legend

—— Nyanyadzi River

NDVI Values

-0.80 - 0.00	0.01 - 0.23
0.24 - 0.30	0.31 - 0.39
0.40 - 0.51	0.52 - 0.85

Figure 6.6a: aerial picture Block C on 21 June 2013

Figure 6.6b: aerial picture Block A on 21 June 2013

Figure 6.6c: NDVI Values for September 12, 2013

Figure 6.6d: NDVI Values for September 15, 2014

Figure 6.6: Satellite images of Nyanyadzi Irrigation Scheme

6.7 Discussion

This article shows that the Nyanyadzi waterscape is constantly reordered as a result of social and natural responses to institutional changes, yet at the same time different socio-political eras have left visible traces on the landscape. Uneven social relations have materialized in the Nyanyadzi waterscape through concrete hydraulic structures, changed environmental conditions and disparate access to land and water resources that "... *long outlive the particular alliances that created them*" (Mosse, 2008: 941). In this way, the current waterscape is an outcome of continuous socio-nature processes in which legacies of former institutional arrangements as well as contemporary interventions shape the physical environment with disparate implications for access to and control over water resources for the water users in the catchment. The responses to the water reform process that were initiated in 1998 also let to a reordering of the waterscape in unexpected ways and with potentially long lasting consequences, both for water users as well as the environments they live in.

The Zimbabwean water reform process has been influenced by a global change in public policies for water resources management in which the main focus of governments shifted from investing in the development, operation and maintenance of water infrastructure to a focus on managing water resources systems by stipulating frameworks for water allocation (Cleaver and Elson, 1995; Allan, 1999; Neubert et al., 2002; Mosse, 2004; Lowndes, 2005; Swatuk, 2005; Saleth and Dinar, 2005; Mosse, 2006; Ahlers and Zwarteveen, 2009; Sehring, 2009; Kemerink et al., *forthcoming*). As such the Zimbabwean government's role within the water sector shifted and they became primarily involved in attempting to steer institutional change processes through defining regulatory frameworks, drafting organizational blueprints and specifying key principles for water allocation. Within this role the Zimbabwean government, realizing the need to redress the racial legacy of the colonial time and in line with the global trend in thinking about water resources management, abolished the priority date principle, riparian right principle and the perpetuity principle for allocating water resources. They reasoned that by abandoning these principles water would become easier accessible for new users, which would include the previously dispossessed indigenous African population. Moreover, the Zimbabwean government recognized the importance of, and inequity in, access to hydraulic infrastructure and therefore established a national fund to stimulate the development of hydraulic infrastructure in previously disadvantaged areas. In contrast to the exclusive focus on institutional change processes that have been adopted elsewhere, this progressive physical oriented measure gave the government the opportunity to directly rearrange water flows and as such affect the distribution of water resources.

Despite this progressive approach of the Zimbabwean government, the reality on the ground is different with water users paying water fees within Nyanyadzi catchment mainly to cover the overhead costs of a heavy organizational structure without receiving improved services and without the much needed investments in hydraulic infrastructure. Moreover, the changed principles for allocating water have affected access to water for the water users in the catchment in unexpected ways. The Nyanyadzi waterscape is quite unique in the country with the oldest right to water belonging to a smallholder irrigation scheme rather than white settler farmers and with the land already transferred to the indigenous African population before the government decided to fast-track the land reform process. Disconnecting water from land by abolishing the riparian right principle just when the indigenous population finally has the legal opportunity to own the land and replacing it by an expensive and cumbersome permit system has forced especially individually operating smallholder farmers upstream in the catchment to illegally access water (see also Manzungu and Machiridza, 2009). Moreover, the abolishment of water entitlements based on priority date has triggered a physical transformation of the Nyanyadzi waterscape; downstream water users have abandoned their hydraulic property and left their land in the irrigation scheme to move further upstream in the catchment to farm in closer proximity of the river where entitlements are perhaps less secure and where sophisticated infrastructure is absent, but access to water is cheaper and more reliable. The bureaucrats involved in crafting the Zimbabwean water reform process did not intend to reduce the water use of lawful permit holders and for sure did not envisage the increase of water use by illegal abstractors, yet this is the outcome of the reforms implemented in the historically produced uneven and contested waterscape of Nyanyadzi catchment (see also Kemerink et al., *forthcoming*). This outcome is not neutral, but highly political since a few influential actors have considerably increased their access to land and water in the catchment, and as such the failure of the water reform process has been used by both the ruling as well as the opposition party in their rivalry campaigns.

Even though the response to the water reform process was not foreseen by the bureaucrats who initiated the reform process, the reordering of the waterscape could with little effort be observed from widely available remote sensing images of the catchment[50]. Such images do not make explicit which changes in a waterscape are the results of a particular policy reform and which changes are caused by other processes, but rather capture relatively easily the physical manifestation of how these complex and dynamic processes of socio-nature unfold on the ground. Therefore these images could aid policy makers who are tasked to achieve particular predefined objectives, but who are faced with ever changing conditions in the catchments in which they intervene. Remotely sensed images may help them to track processes of change and observe how their ambitions on paper unfold within the dynamics of society. As such, it can serve as an important tool for monitoring the outcomes of water reform processes and detecting undesired outcomes. We do not suggest that policy makers should become 'big brothers' who are watching from the comfort of their own office chairs what the people in the catchments do, but rather we suggest that aerial pictures, and more importantly the critical and interdisciplinary analysis[51] of these pictures, can enrich rigorous assessment of policy outcomes and inform the formulation of new policies. This might be a more sensible approach for steering reform processes within dynamic contexts than relying on a set of fixed indicators such as number of permits granted, amount of fees collected and number of meetings held as these say little about the actual distribution and use of water within a catchment.

---

[50] We like to emphasize that, even though extensive databases with satellite images are widely available and freely accessible, access to higher resolution images, longer time series, software packages and skills, all required for more detailed analysis, remain restricted or are only available upon payment of considerable fees.

[51] We emphasize here the need for critical and interdisciplinary analysis since we acknowledge the challenges to combine social science and remote sensing technologies, which has been mainly developed and used within natural science domain (Blumberg and Jacobson, 1997; Rindfuss and Stern, 1998; Macauley, 2009), potentially leading to partial and/or distorted interpretations of the images.

7. Discussion and conclusions: From water reform policies to water resource configurations

7.1 Introduction

"Equity is critical in ensuring that water reform in South Africa is realised ... The existing legal framework and policy does not adequately respond to the objective of redress in terms of making water available and advancing equity considerations. It is imperative that provisions within the National Water Act should not only protect the interest of existing water rights but should also provide mechanism to make water available for redress. There is thus a need to consider how ... the process of redress and achieving equitable allocation of water could be addressed within the policy review process ... that will ensure equitable water allocation and enjoyment of water benefits by all."
(DWA, 2013: 39)

Fifteen years after the endorsing a Water Act that was internationally praised for its progressive and inclusive nature, the South African government rolled out a new water resource strategy, acknowledging in particular that little progress has been made in achieving equitable allocation of water resources. So what happened with that eulogized piece of water legislation? Where did it go wrong? Was, after all, the Act itself not what it pretended to be? Did the government fail to implement its own legislation? Did water users circumvent the reform process or undermine its objectives? How then can equity in water be achieved? This research has been sparked by, and aims to address, these kinds of questions in order to understand water reform processes and how it affects water use.

This last chapter brings together the findings of four case studies on water reform processes in an incorporated comparison and reflects on the broader implications of the research findings for science as well policy practice. A few notes are appropriate at the start of this chapter. First, I chose to write this concluding chapter from the first person perspective to emphasize that it presents my interpretation of the research data and reflects my standpoint. However, by no means this should be interpreted that I pretend to have carried out this research all by myself or in isolation. On the contrary, this research is the result of the efforts of a large team of researchers, supervisors, associated researchers, (anonymously) peer reviewers and journal editors. Moreover, this research is formed by the active engagement with the actors within the four highly dynamic waterscapes in which this research is situated and has been informed by scholars from a wide range of academic disciplines. Second, since I aim in this chapter to move beyond the individual case studies in order to reflect on the broader processes that instigated and shaped the water reforms in the different waterscapes, I will need to generalize the research findings to a certain extent. However, where applicable I will discuss how an issue works out within a particular case study with reference to the relevant chapters for more detailed descriptions. Third, I decided not to include references in this chapter, except of course when I directly quote other authors or present new data, since all theories, concepts and data I use for the analysis in this chapter are extensively referenced in the previous chapters of this dissertation.

This chapter starts with a synopsis of the research both in terms of research approach as well as main findings. This is followed by an incorporated comparison between the four case studies to illuminate the production of social difference within the water reform processes and to discuss the implications for the water resource configurations in the selected waterscapes. Thereafter, I will try to answer the main research question by discussing how, to what extent and why the mainstream approach for public policies has altered institutions that govern water resource configurations. This is followed by an attempt to address the two specific objectives set for this research by discussing the societal and scientific relevance of this research. I will end this chapter with a critical reflection on the choices I made within this research, both in terms of content as well as methodology, and identify topics for possible future research.

7.2 A synopsis of the research findings

This research focuses on understanding water reform processes by studying the institutional processes that steer, stall or tweak the envisioned change within the water realm. Within academia different theories seek to explain processes of institutional change. The mainstream school of thought assumes that institutions can be externally designed and crafted through policy interventions to achieve certain common objectives, while another school of thought, critical institutionalism, argues that institutions emerge from daily interactions among actors with disparate leverage positions who try to pursue multiple, ambiguous and sometimes conflictive objectives. These two disparate theories create a paradox in science on how to understand institutional change processes, and in particular whether or not these processes can be instigated and steered by policy interventions. The empirical research presented in this dissertation shows that institutions that actually govern water resource configurations in the case study areas are the outcome of both conscious external crafting by policy makers as well as the interpretation, negotiation and rearrangement by socially uneven positioned actors with diverse interests. In this process these actors make strategic choices as well as draw on daily routines and existing ways of doing to respond to changing circumstances. How water reform processes exactly unfold depends on the historical social-natural processes that have constituted each of the waterscapes into a unique manifestation. Nevertheless, the global shift in the 1980s towards public policies that focus on institutional rather than physical processes has influenced the water reform processes in the countries studied; even though the political ambitions and implementation strategies are different, the stated policy objectives and the selected policy models for the water reforms show fundamental similarities among the selected countries[52]. As such, the water reforms can be seen as a construct disseminated by a globally operating policy network and this research shows that it produces similar processes of social differentiation within the dissimilar contexts of the case study areas. I am concerned about the implications of this particular policy construct because this research shows that it hampers the access to, control over and distribution of water resources for the majority of agricultural water users within the selected waterscapes, and in some cases even legitimizes the historic inequities of the colonial past despite contrary political ambitions of the national governments. With this research I therefore aim to answer the question to what extent, how and why global trends in the policy making process influence, shape and change the water resource configurations within waterscapes in the hope that this insight will contribute to rethinking the current policy approach for water resources management.

---

[52] Sections 1.2.2 and 1.4.3 discuss the common policy objectives set by the national governments to reform the water sector and Chapter 5 provides a detailed analysis of the rationales used to justify these policy objectives.

This research used the extended case study method to analyse four catchments in different southern African countries that all went through extensive water reform processes during the last two decades. Since the water reforms in the case study countries originate from the same epistemic source and mutually conditioned what emerged as a globally recognizable policy construct, they can be regarded as parts of a larger world-historical process. This has formed the basis for the incorporated comparison between the cases with the aim to understand why this larger process consolidated and how it produces social differentiation within waterscapes (see also Chapter 1). Within each extended case study the implications of the water reform process have been studied by analyzing how it has affected the water resource configurations in the waterscape and what the implications are for the agricultural activities of the water users. Each case focuses on different facets of the reform process in order to thoroughly comprehend the working and implications of the shift in the policy approach that took place since the 1980s (see also Table 1.4 and Textbox 7.1).

Textbox 7.1: Summary of main findings per case study

The case study in **Tanzania** focuses on the negotiations over access to water between and within traditional smallholder irrigation systems. At the time of the research the water reforms had not yet been formally rolled out in the case study area, nevertheless, the government and NGOs had started with interventions with similar objectives such as the establishment of platforms for stakeholder interaction at different spatial levels in the catchment with democratically elected representatives and the introduction of fees to be paid for water allocations. This case study shows the hybrid and dynamic nature of institutions and how these endure, evolve and vanish over time. It gives a detailed account of how water users use different normative frames from various sources to exert their claims in the negotiations over access to and control over water and tells us that traditional, or indigenous, institutions must not be idealized.

The case study area located in **South Africa** illuminates how the reform processes are contested in society and how this shapes the interactions between water users with historically uneven leverage positions. This case study shows that different, sometimes conflicting, normative orders underlie the internationally praised South African Water Act and discusses how this piece of legislation interacts with existing institutions within a still highly segregated society. Moreover, this part of the research shows how the use of seemingly neutral policy models, in this case the decentralization through establishment of water users associations, leads to the reinforcement of structural inequities in terms of access to and control over water resources in the catchment.

The case study in **Kenya** focused on the rationales used to justify the water reform process and unravels to what extent these rationales are valid for various kinds of agricultural water users in the case study catchment. It shows that only a few historically advantaged commercially oriented water users benefited from the new legislation in the study catchment, either by adapting to or by rejecting the water reform process. In particular, this case study identifies several unexpected and undesired outcomes of the reform process for small-scale farmers who are member of water user associations and shows how this is linked with the institutional plurality as well as the type of hydraulic infrastructure these farmers have access to.

The last case located in **Zimbabwe** studies the implications of the implementation of water reform policies within a rapidly changing context due to instability in land tenure and collapse of the national economy. This case study uses landscape analysis to understand the historical development of the waterscape and to show how people respond to the changing conditions, including the water reform process, by reordering their physical environments. Moreover, this case study explores the use of satellite images to incorporate complex socio-nature processes into policy making process to aid policy makers who wish to respond to dynamic and context specific circumstances.

7.3 The emerging water resource configurations

The water reform processes in the case study countries shared, albeit diverging political ambitions, similar policy objectives, namely; to provide security in access to water users; to decentralize responsibilities and include water users in decision-making processes; and to encourage efficient use of water by charging fees for service provision. The empirical data presented in this dissertation shows that these objectives have only been achieved for particular groups of water users in each of the case study areas. For other groups of water users these objectives have only partially been achieved or even led to disparate outcomes. None of these groups of water users are homogenous nor are they similar across the case studies per se. However, these particular groups of water users do share comparable accounts in terms of historically produced social identities, subjective social relations and material (dis)possessions. I thus observe that the water reforms have contributed to similar processes of social differentiation that have shaped the water resource configurations within each of the study catchments.

The water reforms have led to improved water security for large-scale farmers who hold an individual permit to use water and have (at least partly) individual control over hydraulic infrastructure. This group of water users, which in this research consists of farmers who established their enterprises during the colonial occupation and/or obtained water rights under previous legislation, also have been successful to securing a substantial voice in the decision making platforms. Most of them have been involved from an early stage in the implementation of the reform process and could easily adapt their practices to the reform process because they had past experiences on which they could draw. For example, most large-scale farmers in the South African case study are for several generations member of State-sanctioned irrigation boards that have a similar kind of organizational structure and comparable bureaucratic procedures for decision-making as the decision making platforms established under the reform process (Chapter 4 and Annex 2). And in the Kenyan case study the large-scale farmers had experience with the cumbersome administrative procedures required for obtaining water use permits under previous legislation and they obtained advice from large-scale farmers in adjacent catchments that were ahead with the implementation of the reform process when they took the initiative to establish a water user association in their catchment (Chapter 5). This familiarity and know-how gave the large-scale farmers an advantaged position within the water reform process and as such they could increase their control over the water resources in the catchments, not only securing their current but also their future access to water. Nevertheless, these farmers pay a price for these advantages as under the water reforms fees have been introduced based on the amount of water allocated. Even though their businesses rely heavily on loans from investment banks, by having access to (international) markets they can sell their products, such as grains, vegetables, dairy and flowers, with profit and reinvest in modern infrastructure and new technologies that helps them to sustain their advantaged position within the waterscapes.

The cost for accessing water also has increased for farmers who farm within smallholder irrigation schemes. However, despite having collective water permits in some cases, their water security at farm level has not improved due to various reasons including inadequate hydraulic infrastructure to distribute and store water, sharing water among an increasing number of farmers who rely on the irrigation scheme and managerial issues. Moreover, these farmers do not feel included nor represented within the bureaucratic decision-making platforms; the issues they face are not on the agenda nor are the procedures set for selecting representatives for these platforms and the rules for interaction aligned with the institutions embedded in their daily realities. This group concerns the majority of water users in all four case study areas and mainly consists of small-scale farmers who grow crops for subsistence and only sell when they have a good harvest. Whereas in most cases the necessity to collaborate among these farmers was initiated by a physical imperative in terms of the collective action needed to construct and maintain the shared hydraulic infrastructure, it seem to have shifted under the reform process to an administrative imperative to obtain (and pay for) a water use permit and to claim voice in the newly established decision-making platforms. It is interesting to note that even where progressive measures were taken within the reform process to protect the interests of small-scale farmers, those who rely on collective access to natural resources and/or infrastructure have not profited from these legal instruments. For example, the farmers in the smallholder irrigations schemes in the case study in Kenya end up paying proportionally more for their collective water permit than they should based on their individual water use (Chapter 5); despite the redistribution rhetoric in the South African Water Act, the farmers on communal land in the case in South Africa struggle to allocate suitable land for constructing a dam to store water and thus continue to see the water flow off their lands into the downstream reservoirs of a large-scale farmer (Chapter 3); and farmers in the irrigation scheme in the case in Zimbabwe have moved from being the first in the water queue to being the last, detrimentally affecting their water availability, because it is them who lost their prior date right to the water under the reform process rather than the large-scale commercial farmers this clause was meant for (Chapter 6).

The exceptions to the rule are those small-scale farmers who manage to 'escape' the reforms by illegally accessing water, either by unlicensed abstracting water straight from the source, for example in the Kenyan and Zimbabwean case studies (Chapters 5 and 6), or utilizing patronage systems and/or bribery to acquire (additional) water from a collective water source, for instance in the Tanzanian and Zimbabwean case studies (Chapters 2 and 6). Often these are people who historically have a privileged position within the waterscape, either in terms of physically advantage such as owning land in close proximity to a river or dam or because they are socially well connected to the local elite and/or the ruling party and receive protection and material gain through these channels. Even though these practices have always been a source of conflict among water users, to some extent it was also met with acceptance by other users and as such was socially embedded in the institutions that govern water resource use in the catchments. Despite the fact that the water security for this group has weakened as their water use is now labelled by the government as illegal, and even though they are not represented in State-sanctioned platforms at catchment level, their physical access to water is often better than the other small-scale farmers, allowing them to grow more crops to sell on local markets. Moreover, these farmers avoid paying fees for the water they use, except for small monetary or in kind bribes in some cases to satisfy their patrons. The money that they earn this way can in turn be invested in pumps or small hydraulic structures increasing their access to and/or storage of water.

In addition to the groups discussed above, in all four catchments there are subsistence farmers who rely solely on rainfall as a source of fresh water for growing their crops. The water reform processes have changed little for them especially because there is little attention within the policy documents on investing in the development of hydraulic infrastructure to increase their (green) water uptake. This becomes evident from the drop in public as well as private investments to support irrigated agriculture in sub-Saharan Africa since the shift in the public policy approach has been introduced in the 1980s (see also Chapter 1).

The generic, decontextualized outcome of the reform processes in the four case study countries is thus that it contributes to processes of social differentiation that mainly benefits historically advantaged water users who, at least partly, have individual control over access to water and who produce their crops primarily for the commercial market. With a few exceptions (see for example Chapters 2 and 5), it should be noted that within the study catchments the outcomes of the water reform processes are therefore largely skewed along racial lines since the historically advantaged large-scale commercial farmers are from European descent, while the marginalized small-scale farmers have indigenous African roots. Moreover, the reforms have gendered implications within the researched waterscapes since those who managed to tweak the implementation process in their favour are primarily male farmers. Based on the above I conclude that in all four cases the water policy interventions have changed the water resource configurations within waterscapes studied under this research but in a particular yet limited way. It seems that institutions can thus, at least partially, be crafted through policy interventions. However, the question remains *to what extent* and *how* this happens and, perhaps more importantly, *why* the mainstream approach in water policy reforms led to these particular outcomes? Let me start with answering the question *how* the mainstream public policies interact with and alters institutions that govern access to and control over water resources within waterscapes.

7.4 Policies lost in translation?

In none of the case study areas the policy objectives have been fully achieved and in most cases it has sparked unexpected developments with sometimes adverse outcomes. This points to a more complex and dynamic process than straightforwardly implementing a public policy and enforcing externally designed rules. For this research I have been particularly interested in illuminating what happens between the government's policies on paper and the water resource configurations and management practices that emerge within the waterscapes. The interactions over water between the farmers in the case study areas show that the institutions governing the water resource configurations are dynamic in nature, constantly negotiated, reconfirmed and contested. In this process actors actively use the normative frames and institutional blueprints that have been introduced by the national governments as part of the water reform process. They, consciously and unconsciously, have interpreted, reworked, adopted and rejected parts of the government's policies and combined them with existing institutions into new hybrid institutions. In this process not only the policy itself but also the approaches and instruments selected for the implementation of the paper policy play a role in determining the outcomes of this dynamic process. For instance, the choice to use existing white dominated irrigation boards as a starting point to establish racially mixed water user associations in South Africa has greatly compromised the inclusiveness of these associations (Chapter 4) and the external pressure to use quotas for appointing women in the water management committees did little to address the structural causes of gender inequity in the case study in Tanzania (Chapter 2). Moreover, government officials tasked to facilitate the implementation of water reform processes at local level are actively involved in framing and interpreting the policies according to their own perspectives and experiences and as such steer the translation of the policies from paper to the local reality within the waterscapes. For example, a government official in South Africa explained the guidelines from the Department of Water Affairs, stipulating the need for a *"balanced representation in terms of the various categories of users"* (RSA, 1999:18), in such as way that it boiled down to the color of the skin of the water users rather than the purpose of their water use (Chapter 4).

As discussed in the first chapter of this dissertation, the agency of actors is neither rigid nor equal, disparately circumscribing their capability to respond to and manipulate policy interventions. The accumulation of wealth through dispossession under colonial rule has led to a highly uneven distribution of material resources, hydraulic infrastructure and capacities in each of the case study areas. Within these uneven waterscapes historically advantaged water users have been able to exercise a stronger influence on the processes of bricolage through which the institutions have materialized, evolved and endured. They could manipulate interactions over water in particular ways so that the institutions that emerged from this negotiated process best served, or least harmed, their interests. This also happened within the water reform processes staged within these physical landscapes that have been constituted by (responses to) former normative frames and legislation. In the case study area in Kenya as well as in South Africa, for example, the historically advantaged large-scale farmers have tweaked the reform processes in such a way that it did not only increase their own security to water, but also gave them instruments to restrict the water use of other, less advantaged, users within the catchments by claiming the hydrological boundaries as 'natural' jurisdiction of the newly established water user associations (Chapters 4 and 5). Also within the Zimbabwean case study historically advantaged users manage to keep control over the agenda of the new collaborative platform at river basin level; councillors that represent the large-scale commercial farming sector continue, perhaps rightfully, to express concerns and ask questions about administrative and financial issues, redirecting the council away from tangible activities directly related to water resources development and management, and stalling discussions on the more contentious issue of water allocations within the basin. This navel-gazing is also reflected in the annual budget of the council which is largely spent on keeping the organization running (e.g. staff salaries, fancy meeting venues, offices and allowances for councillors) and hardly on implementing the council's core activities of regulating water use and supervising the day-to-day management of the water resources (Chapter 6).

Nevertheless, as discussed in Chapter 1, hegemony is never absolute and thus historically disadvantaged water users also have agency. These water users as well use the water reform process as an opportunity to contest established authority and renegotiate existing institutions that govern water resources. For instance, the members of the clan who established a furrow irrigation system in the case study area in Tanzania saw their priority use of the water diminishing due to the enforcement of democratic principles in the management of these indigenous irrigation systems (Chapter 2). Moreover, the democratization of the management of these systems triggered an increase in the number of farmers relying on a downstream system so that they could obtain majority vote in the negotiations with upstream, more advantageously positioned, irrigation systems in the negotiations over water. In the case study area in Zimbabwe farmers reordered the waterscape in response to the reform by abandoning their land in the irrigation system and moving their agricultural activities upstream (Chapter 6). Even the mundane choices of not attending meetings of the water user associations in the case study in South Africa (Chapter 4), not submitting applications for water permits for individual small-scale abstractions from the river in the case study in Kenya (Chapter 5) and irrigating the fields in the middle of the night in the case study in Zimbabwe (Chapter 6) show the agency of the small-scale water users in their refusal to be incorporated in the reform process.

Moreover, the water users are not only connected to each other via the water that flows through the landscape, but they also form relationships based on use of other (natural) resources, vicinity, family ties, employment, clientele, race, ethnicity, gender, religion, political association, nationality and so on. Within these complex webs of affiliations subjectivities are achieved and ambiguous social identities emerge in which somebody is not just a water user, but also for instance a single mother, an employee, a neighbour, a pastoralist, a fellow believer and a black feminist activist. As result of these subjective yet interdependent social relations actors do not solely strive for optimal use of water resources, but actors might also pursue other objectives, including maintaining or contesting these same social relation. For instance, in the case in South Africa the large-scale farmer that resides directly downstream of a former homeland to some extent accepts damage to his property, including destroying expensive irrigation equipment and stealing of crops, because he wants to sustain a peaceful relationship with his neighbours as they are also his employees, customers and the protégées of his political opponents. Also when it comes to water itself relationships are not straightforward: the same large-scale farmer, who seems to have the sole right to the water in the small sub-catchment, nevertheless depends on the people residing in the former homeland upstream as they can easily manipulate the water he relies on by blocking the water flow or increasing the sediment load, silting up his reservoirs (Chapter 3). In the case in Tanzania the smallholder farmers in a downstream irrigation system fail to negotiate more advantaged water sharing arrangements with upstream irrigation systems partially because of family ties and political affiliations between farmers in the systems, while the same kinship also provides security in farming during droughts: instead of water flowing downstream, people move upstream to cultivate on plots of relatives and associated farmers. Moreover, disparate access to water resources, in combination with different soil types in the catchment, does not only create conflict, but also leads to diversity in the crops grown, stimulating trading among, and thus creating customer relationships between, the farmers (Chapter 2). These ambiguous and contested social identities make categorizations like 'community' or 'emerging farmers' or 'the poor' for policy and/or research purposes problematic, because these labels do not capture the real complexity of somebody's social identity nor reflect their everyday struggles. Hence, these categorizations tend to reinforce subjectivities and as such produce social differences among actors.

Based on the above, I conclude that, rather than through externally designed crafting processes steered by policy makers, the water reform policies have altered the institutions that govern the water resource configurations through *complex* and *uneven* processes of bricolage. In these processes not only water users but also government officials actively participate, trying to manipulate institutions in an attempt to not only pursue the stated and unstated policy objectives but also to suit their own understandings and interests. Once enacted, policies thus add to the legal repertoire socially uneven positioned actors can draw on in a continuous bargaining process to establish the institutions that determine access to, control over and distribution of water resources. This has led to unintended outcomes and thus to a disjuncture between what is written on paper and what emerges within the waterscapes. I therefore argue that the policies have to some extent been lost in translation somewhere within the implementation process. However, the similarity in the outcome of the reform processes in terms of water resource configurations points to a more structural rather than random process.

To understand to what extent and in which direction the water reform policies have altered the institutions I find it useful to use the metaphor of institutional corridors (see section 1.2.3). Despite differences in the colonial and post-colonial history of the case study countries, the colonial occupation has led to the proliferation and persistence of particular norms in each of the societies and as such narrowed the institutional corridor by restricting the legal plurality and limiting the actors involved in decision making. Evoking change through reform processes in such a constricted setting is often difficult as established interests in society create institutional stickiness. Especially in the catchments in Kenya and South Africa, where large-scale farmers from European descent are still operating their businesses, the institutional corridors seem narrow with little room for manoeuvre for other actors. Even though also this group is not homogeneous in many ways, they have built a strong collective identity and are well organized among themselves. Most of these farmers are familiar with the rhetoric of the policies and have been able to aptly mobilize their assets and networks to tweak particular interpretations into the reform process and as such steer the process in a certain direction. For example, the large-scale farmers in the South African case study have used the economic meltdown in neighbouring Zimbabwe as an argument to discourage small-scale farmers to claim redistribution of land and water resources (Chapter 3) and the large-scale farmers in the Kenyan case study have spun a particular interpretation into the reform process that forces small-scale users to become member of a water user association if they wish to use water from the river (Chapter 5). Consequently, only those articles of the water reform policies that serve the interests of the few historically advantaged water users have been 'aggregated' by adopting and combining them with the existing institutions, but not necessarily changing the essence of these institutions, while the implementation of articles of the policies that are less beneficial to them have been stalled. In this way they have been able to use the reform process to their advantage to reinforce their control over the water resources in the waterscapes and further narrowing, rather than widening, the institutional corridor. Even though bricolage is a legitimizing process, it should be noted that it is not merely a consciously steered process in which actors make solely strategic choices. Rather, it is also partly an unconscious process in which actors draw on existing practices and familiar institutions to improvise in response to changing circumstances. For instance, the South-African government has given little practical guidance and support to the large-scale farmers on how to involve marginalized and potential future water users in the existing platforms on water management and as such fulfil their legal obligation. These farmers thus had to be inventive and rely on their own understanding and experiences to deal with the complex process of including thee excluded (Chapter 4). Understanding power as a Foucauldian notion in which it is not only possessed and exercised by actors, but also operates through the existence and internalization of structurally uneven institutions, and understanding waterscapes as historically constituted by socio-nature processes (see also Chapter 1), I conclude that (parts of) the policies that share normative underpinnings with the prevailing institutions in society are easier adopted, while (parts of) policy reforms based on alternative normative views have less chance to materialize and affect water resource configurations. Since the reforms processes have altered the water resource configurations in the case study areas, the global policy construct must, at least partially, have resonated with normative frames upheld within the societies studied under this research.

The question remains *why* the water reform policies in four different countries were framed within normative understandings that aligned with those of historically advantaged actors, allowing them to strengthen their position within the waterscapes, despite progressive political ambitions to redress the colonial legacy, at least in South Africa and Zimbabwe.

7.5 Connecting policies with the outcomes

The institutional corridor has not only been narrow within the uneven waterscapes, but also at national level in the case study countries. The colonial history has left the countries with limited human resources within government departments and a particular epistemological positivist legacy that favours technocratic and as such depoliticized approaches. This, together with a high dependency on donor funding[53], made policy making processes at national level susceptible for blueprints circulating within global operating policy networks. As such, even though the political ambitions differed, the water reform processes in all four case study countries have to a large extent been instigated by the change in thinking about water resources management among international experts that fused with a global shift towards implementing a neoliberal agenda for the delivery of public services that obtained momentum in the 1980s (see also Chapter 1). This accumulated in the adoption of the mainstream approach for public policy in the case study countries that pushed the content of water policies from a physical orientation to a focus on steering institutional processes. The water reform policies are thus largely constructs produced by a particular epistemic community within a decontextualized setting rather than an outcome of rigorous formulation processes at the national levels in which policy narratives are verified and policy models are scrutinized.

The particular policy model that was rolled out under these reform processes disconnected water from land resources and aimed, amongst others, to secure access to water for (private) entities through compulsory licensing of water use via permits stipulating the amount, the duration, the conditions and the (single) purpose of the water use; to delegate management responsibilities to semi-autonomous organizations at catchment level that operate on bureaucratic and democratic grounds; and to charge monetary fees based on the amount of water used to encourage efficient use and to recover (part of) the costs for managing water resources. The emphasis in the policy documents on steering reforms through crafting institutional change processes circumscribed the implementation instruments available to the governments primarily to legal and financial interventions. Even though some of these policy instruments aimed to redress the colonial inequities and/or protect the interests of small-scale water users, the water reforms largely followed a neoliberal normative frame that catered for market oriented producers who have access to hydraulic infrastructure that allow them to rigorously control the water flows, excluding the far majority of citizens who rely on communally owned rustic infrastructure that does not allow for full control of water or on rainfed subsistence farming. The political nature of the marginalizing process initiated by the adoption of this policy model has been concealed by 'progressive' indicators, or perhaps I should rather use the term vindicators, like the number of blacks in executive positions of the water user associations, the number of women attending water allocation meetings, the number of permits granted to communal smallholder irrigation systems or the amount of water fees 'willingly' paid by small-scale farmers, but leaving out the most relevant, and thus most political, barometer, namely the actual (re)distribution of water and water related incomes in society.

It is interesting to note that, paradoxically, even though the water reform policies have been inspired by integration rhetoric, it detaches the social from the physical by disconnecting water from land resources and by beforehand excluding technological policy instruments such as the investment in hydraulic infrastructure, even if these are needed to achieve the policy objectives. For instances, the small-scale farmers in Tanzania cannot use water more efficiently since up to 80% of the water leaks from the earthen furrows (Chapter 2); the small-scale farmers in the Kenyan case study cannot use water more productively since there is not enough storage in the irrigation scheme to adjust the water supply to the specific water

---

[53] Even though South Africa received no official development assistance (ODA) during apartheid and only a very limited amount since 1995 (about 0.2 to 0.4 % of GDP of which 2.1% is spent on agriculture) most ODA is geared towards technical cooperation through which consultants from donor countries are hired to support national policy development (Ramkolowan and Stern, 2009).

requirements of their crops (Chapter 5); and the residents in the former homeland studied in South Africa cannot meaningfully participate in the WUA because, amongst others, the limited land they have access to and the absence of hydraulic infrastructure prevents them from using 'relevant' quantities of water (Chapter 3). With this I do not argue that land reforms or (plans for) infrastructural development will not be manipulated and affected by uneven processes of bricolage(see for examples Chapters 4 and 6 and Annex 2), and therefore I do not claim that by linking water and land reforms and including technological instruments the water policy objectives will be achieved, but rather I want to point out that without targeted government investments in access to land and infrastructure the policy objectives can certainly not be achieved for the majority of the agricultural water users in the case study countries. This research thus shows a disjuncture between the policy objectives and the selected policy instruments to achieve these objectives since large parts of the water legislation enacted under the reform processes is not attainable for the majority of the agricultural water users because they lack access to land and (adequate) hydraulic infrastructure.

Especially excluding investments in the development of hydraulic infrastructure for historically disadvantaged groups has severely narrowed the options and the capacity of the governments to redress the colonial legacy. These targeted investments could open progressive trajectories for water reuse and redistribution that otherwise most certainly remain impossible, thus leaving the smallholder farmers with little chance to increase their water use and move their livelihood beyond subsistence. The politicized rationale of the choice to exclude these technological policy instruments becomes apparent by considering history; during colonial rule the uneven waterscapes have been produced largely through government investments in land acquisition and development of hydraulic infrastructure[54] based on racial differentiation as well as protection of markets. At that time commercial oriented, individually operating actors from European descent could thus establish their businesses with substantial support from the government. Now these same actors (or their heirs) benefit from the institutional reforms *thanks to* the reduced role of the governments and *thanks to* the neoliberal notions embedded within the selected policy model emphasizing the need for efficient and productive use of water. In other words, there is a direct link between the outcomes of the water reform processes and the partial focus on, and normative framing of, the institutional instruments selected for the reforms. Not only does this reinforce, and even legitimize, the inequities of the colonial past in terms of water resource configurations, it also questions the role of national governments in safeguarding the interests of society as a whole and the interests of vulnerable groups in particular. Policies are not only an input into processes of institutional bricolage but are also an outcome of similar processes and as such policy making is equally circumscribed by uneven social relations established along the global-local continuum. National governments thus do not have full control within policy arenas, yet perhaps even more worrying, they seem to have internalized market mechanisms as the norm for distributing water and water related rights, risks, responsibilities and income. Also here the Foucauldian notion on power is useful to understand how uneven relations of power have materialized into hegemonic normative frames and as such are embedded in broader forms of social, cultural and economic structures. This links to academic debates on governmentality (Foucault, 1979; 2000a) and governance-beyond-the-state (Swyngedouw, 2005; 2011; Jessop, 1998) that discusses not only the 'rules of the game' but also the 'game of

---

[54] Even though also the post-colonial governments made considerable investments in infrastructure including large dams for hydropower and water supply to urban areas, and as such contributed to the transformation of the waterscapes, investment in irrigated agriculture remained limited in the case study countries, and in particular in the case study areas, and was mainly geared towards rehabilitation of the existing schemes (Faurès et al., 2007).

the rules' and the particular role of governments herein. These discourses are not specific for the case study countries studied within this research, rather these governments need to operate within a perhaps even more intricate governance processes as result of the colonial and post-colonial past that contributed to structural inequities in society and made these governments dependent on external finance and expertise for policy interventions.

I can only conclude that policies do have agency within waterscapes, perhaps not in the sense that they have the ability to act autonomously, but rather in the sense that, once ratified, public policies are not easily replaced or abandoned and as such they gain impetus, opening particular trajectories while closing others. Especially when policies are aligned with the interests of the elite and rolled out through seemingly neutral or even 'progressive' policy models, this research has shown that they have the ability to shape water resource configurations beyond time and space and beyond 'good intentions', affecting dissimilar waterscapes in similar ways. For instance, those government officials and water experts who did have the genuine ambition to redress inequity within the water domain did most likely not foresee that the paradigm shift to an integrated approach for water management would so easily be hijacked by the adepts of the neoliberal mainstream, leading to these particular outcomes. And even now, when governments start to realize that their progressive ambitions are not yet met under the water reform process and try to reformulate implementation strategies, like South Africa recently did, it proves difficult to change the course of the reforms as the predisposition is deeply rooted in the policy model on which these reforms are built. I must therefore also conclude that policies only to a limited extent can contribute to progressive societal change, especially in this neoliberal era as the interests of influential actors operating within national and international policy arenas are so tied up and fixed within a particular normative understanding of the world that large parts of society almost seem to have sanctioned through mundane processes of *"disapproval, criticism or simply an absence of response"* (Giddens, 1984: 175). It is therefore time for us critically oriented scientists not only to take sides and critique the contemporary policy making processes, but also to 'captivate' policy makers and to engage with their daily reality by providing policy alternatives, at least in an attempt to push for a different political trajectory in society.

7.6 Contribution to policy practice

How can those many words in this dissertation captivate policy makers? What is the value of emphasizing the complexity of everyday life when the reality of policy makers is that they can only work with simplifications thereof? And how can the findings of this research be understood in terms of alternatives for current policy practices? What this research shows is that progressive change is not easily evoked through policy interventions since societal transformation is a messy process circumscribed by structural uneven social relations. Nevertheless, yet perhaps naively acknowledging the bounded and contested role of governments in contemporary policy making processes, I would like to suggest three points that could help in revisiting the current policies within the agricultural water realm in the hope it will contribute to redressing historical inequities, namely:

1) The *'political'* needs to be brought into the policy making process. This refers to embracing the political nature of reform processes and making this explicit within the formulation, implementation and evaluation of public policies. This includes, amongst others, a need for a more profound and interdisciplinary understanding of policy issues, explicitly stating the assumptions made for the required simplification of

reality, making explicit political choices and formulating realistic policy objectives, dissecting biased policy models and their origins, carefully selecting policy instruments and implementation approaches, and monitoring and adjusting policy interventions by measuring the objectives rather than the means of the policy reform. This is perhaps difficult to achieve within the existing status-quo as it requires sufficient human resources, both in terms of quantity and quality, within national and local government agencies, something in itself that is highly political. Since the epistemologies and interests of scientists and private consultants are not necessarily aligned with those of governments, outsourcing these activities is also problematic. Perhaps a starting point could be to focus on the next generation by revising educational programmes so that graduates, including future government employees, are better prepared to guide, monitor and respond to the formulation and implementation of reform processes. For water related programmes this could for instance entail that, beyond teaching subjects related to physics, engineering and planning, students will be exposed to the fundamentals of political sciences, learn how to critically evaluate governance frameworks, practise policy analysis methods and develop conflict mediation skills with the emphasis on social inclusion. This requires breaking away from a positivist epistemology still dominant in most water related programmes and focusing on nurturing critical thinkers, capable to reflect on their own viewpoints and practices.

2) The *'context'* needs to be brought into the policy making process. This starts with recognizing that policies, just like institutions, are the outcome of uneven processes of bricolage in which existing arrangements and styles of thinking from other domains or other localities are pieced together into 'new' policy documents. In other words, policies are not 'holy grails' but build on vested interests and ad-hoc improvisation and as such might, or might not, or might partly, achieve the set objectives. Moreover, policy on paper requires a generalized and simplified model of reality, while they are implemented within comprehensive, diverse and continuously changing contexts. Reforms can thus not be obtained through single and straightforward policy interventions but require profound processes of trial and adjustment. This means space is needed within the implementation processes to engage with the multifaceted, plural and contested nature of society and requires flexible and sensitive approaches that are guided by, and respond to, actual outcomes (e.g. the distribution of water and water related incomes) rather than lists of predefined tortuous indicators. It also requires policy makers and implementers who are aware of their role as bricoleurs, actively using both policy rhetoric and existing institutions if and where appropriate and critically reflecting on their own practices. Considering socially embedded institutions and practices is especially crucial when engaging water users in decision making processes, but it also applies for other policy interventions. After all, injudiciously enforcing water use permits and payments for water use, because the policy narratives appear coherent and consistent on paper, might not be the best implementation strategy. Instead policy makers could scrutinize how these measures unfold within a particular catchment depending on the existing water resource configurations and then assess if licensing and/or charging fees accomplish what they aimed for. The necessity for this modest approach becomes even more evident realizing that in most catchments policy makers know little about the actual available water and the amount of water already in use. Furthermore, such a dynamic and context-sensitive approach requires policies that go beyond empty buzzwords and policy makers who are cautious with using dichotomized demographic categories such as rich/poor, man/women, commercial/subsistence, black/white, irrigators/rainfed farmers, and urban/rural.

Within the agricultural water realm tracking physical changes in waterscapes through widely available satellite images and critically analyzing the causes and implications of these transformations can aid policy makers to understand how water users respond to the reform processes and how it affects the water resource configurations.

3) The *'physical'* needs to be brought (back) into the policy making process. This refers to recognizing the agency of non-human nature, including ecological processes and hydraulic infrastructures, in shaping policy outcomes. This starts with the need to gain more knowledge of the physical environment in which the policy interventions will take place, amongst others the availability of and variability in water resources, the main soil properties, the state of the aquatic ecosystems and the capacities and locations of dams and water intake structures. This might entail investment in studies to acquire this data and monitor changes during the policy implementation. Remote sensing might be useful for assessing the water resources available within ungauged catchments. Furthermore, it would require not treating water resources in isolation but explicitly linking it with other natural resources and spatial planning processes in general. This calls for coherent strategies and multi-sectoral management structures across policy domains such as land tenure, water, agriculture, forestry, environment and spatial planning. For example, it would mean integrating water, land and agrarian reform policies and discontinuing the establishment of new platforms that are geared towards dealing with a single resource such as water users associations. And perhaps more importantly, it means shifting away from the neoliberal inclined mainstream public policy approach with its partial focus on institutional processes towards a more comprehensive and inclusive approach that, amongst others, incorporates technological policy instruments such as government investment in, or subsidies for, the development of hydraulic infrastructure for marginalized groups, especially in countries that need to redress a colonial legacy.

7.7 Contribution to theories, concepts and methodology

So what does this research brings that we do not already know? And what is the value of this research for scientific inquiry in general? It is always difficult to assess the impact of a particular study and, especially with the fast developing scientific knowledge base, supplemented by so many publications every single day, it is impossible to keep track on what conclusions other scientists reach. It is therefore important to remain modest and accept that this research effort might not be groundbreaking but rather adds to an existing field of study. In this section I try to reflect on what I consider as the scientific contribution of this research.

First, this research made a serious attempt to carry out interdisciplinary research in terms of drawing on diverse theories, employing multidisciplinary research methodologies and analyzing social as well as technical empirical data. Even though the need for interdisciplinary studies is nowadays well recognized, truly interdisciplinary research is still scarce. Although this research did not produce a ready to use research framework that can be taken up by other scientists, it does give a clear example of how an interdisciplinary concept as socio-nature can be utilized for policy analysis by simultaneously studying how actors interpret, strategize and respond to water reform processes as well as how this affect, and is affected by, where the water flows through the waterscape. Second, this research provides, as one of the few studies, empirical work on, and comparison between, water governance processes in four different African countries, each with their own specific institutions, contexts and pathways in relation to colonial and post-colonial conditions. I have done so by carrying out multiple extended case studies to explore the outcomes of interconnected albeit disparate water reform processes within the countries. Third, within the case studies I analyzed the implications of, and interaction between, the three main policy prescriptions that guided the reform processes, namely delegation of water management responsibilities to formalized groups of water users, the introduction of conditional water use permits for private entities and the payment of fees for water use. By studying these three policy prescriptions in conjunction, this research contributes to advance the in my view much needed understanding of the interplay between concurrently operating dimensions of neoliberalization within water resources management. Fourth, this research has used, critically assessed and/or further developed the concepts of hydro-solidarity (Chapter 2), legal pluralism (Chapters 2 and 3), policy narratives (Chapter 5) and waterscapes (Chapters 5 and 6). Moreover, I have introduced a new concept, namely water resource configuration that I define as *"the materialized division in control over, access to and distribution of water between water users sharing the same water resource"* to emphasize not only the social but also the historical and physical nature of the process through which water resource configurations are produced and maintained (Chapter 1).

This research draws extensively on the emerging theory that can be referred to as critical institutionalism. Based on literature review as well as through analysis of the empirical data this research has contributed to enrich this theory in four ways. First, I have included what I call the *'agency of policy'*. Critical institutionalism is well advanced in analyzing the emergence and existence of contextualized local institutional arrangements as well as how they affect different actors differently, though it pays less attention to structural configurations of institutional processes at larger spatial scales and how these configurations interact with the institutional arrangements at local level. To include this local-global continuum I have employed the extended case study method to analyze case specific outcomes vis-à-vis structuring processes at national and international level as well as across different historic political eras. Moreover, through the incorporated comparison the case studies within this research are brought together as parts of a larger world-historical process in order to understand, and give substance to, this structuring process. From a theoretical perspective I have explicitly linked critical institutionalism to political theories on contemporary policy making processes that explain the persistence of policy models and the discursive practices of networks of policy bricoleurs at various spatial scales within the water realm. This allowed me to untangle the agency of policies[55] within institutional change processes and, in particular, to unravel the pervasiveness of neoliberal shifts despite the context specificities of how they unfold at local level (Chapters 4, 5 and 6). Second, I have included what I refer to as the *'agency of the physical'*. Although many scholars that I consider part of the critical institutionalist school of thought emphasize how the social constitutes the physical, few scholars discuss how the physical constitutes the social. I have done so by employing theories on socio-nature and conceptualizing waterscapes as dialectically produced by unevenly positioned actors of human and non-human nature in an ever ongoing process. In this I particularly looked at the agency of hydraulic infrastructure as well as the materiality of water in shaping institutions that govern water resource configurations within waterscapes (Chapters 2, 5 and 6). Third, I have included the *'agency of the invisible'* by drawing on a Foucauldian notion of power to analyze in which ways ideologies and normative views shape human agency in the interactions over water. For this purpose I studied the normative perspectives underlying policy interventions as well as the normative orders that prevail is society and as such have materialized within landscapes. Moreover, I analyzed how policies contribute to

---

[55] See Section 7.5 for a more detailed discussion on how I define the agency of policies.

establishing subjective relations among various water users and shape their social identities (Chapters 2, 3 and 5). Fourth, this research has contributed to advance critical institutionalism by including what I refer to as the *'agency of applicability'*. If theories remain purely hypothetical without any practical application I believe they sooner or later cease to exist, while theories that are picked up by 'practitioners'[56] often receive momentum as mainstream institutionalism has shown[57]. By attempting to reach out to policy makers, through engaging with their daily reality and providing them with concrete directions for revisiting the water policies, I aimed to make critical institutionalism more relevant for the policy making practice (Section 7.6).

7.8 Further research

As any scientific research, this research raises questions that could guide future studies. In particular, I believe there is a need for more ethnographic studies on the actors involved in policy networks and how their interests relate to the policy models they disseminate. This kind of research would not only include studies on how policy entrepreneurs within the global networks frame problems and disseminate policy models, and how these are linked to their worldviews and tied up to (personal) interests, but also the role of national and local government officials in these processes. It would be interesting to understand better how government officials interpret and use these policy models, how these shape their interventions, and how their (routine) practices potentially contribute to the 'conduct of the conduct'. This research could be done by 'following' policy makers while engaging in policy formulation processes, while interacting with society when they introduce new policies, and while evaluating and reporting to parliament to account for their interventions. Such research would be much broader than only water related policies, though the water domain might offer an interesting setting since it is influenced by a strong global policy network and often water policies affect large parts of society and reorders physical environments. Moreover, it might be interesting to use this approach to study development aid policies since these policies have an additional dimension when it comes to the 'politics of impact' in the sense that the policy makers are accountable to citizens of one country while affecting the lives of people in another country.

In addition, further policy research could focus on detailed incorporated comparisons between interconnected policy processes in dissimilar locations, especially paying attention to how these process-instances are mutually conditioning and affect global policy constructs, and as such constitute each other.

Another research line that in my view could be further strengthened relates to the agency of (hydraulic) infrastructure and the materiality of water, and other natural resources, in order to understand how these shape social relations. This kind of research could focus on studying how physical processes and artefacts produce relations and how these affect the interactions

---

[56] I choose to place the word practitioners between quotation marks, because it is a rather vague term. After all, researchers also practice a particular practice. However, what I refer to here is theories that find an application beyond scientific inquiry by being ratified by influential actors outside academia.

[57] Also here I could draw on Foucault's work, in particular Power/Knowledge (1980), in which he discusses the production of knowledge as an integral part of struggles over power by arguing that for something to be established as a fact or true, it has to be sanctioned by those in positions of authority. In these processes other equally valid statements are denied or discredited (Mills, 2003).

between actors and circumscribe their actions. In this research I looked at how the nature of water, in the sense of flowing downstream and carrying sediments, affects collaboration over water between upstream and downstream users. This understanding could be further enhanced yet this research line could also include studies on the materiality of water from other sources, for instance actors exploiting less visible resources such as groundwater or studies to understand how ownership of temporary water ponds is interpreted and negotiated. Moreover, research could be done to better understand the role of infrastructure in constituting socio-nature processes, for instance carrying out more detailed studies than provided in this dissertation on how the choice for (or absence of) decentralized water storage shape social relations or how the modernization of collectively owned hydraulic infrastructure affect institutions that govern access to, distribution of and control over water. From a complete different angle, this research could also look at how disasters such as large floods or long droughts affect institutional processes; what happens when relatively gradual processes of bricolage, in which actors improvise and negotiate to deal with daily challenges, are disrupted by such devastating events? Does this simply lead to a fast-tracking of bricolage or do complete other processes affect the institutions that emerge?

In general, it would be good to further investigate the potential and pitfalls for using remote sensing techniques for enhancing critical analysis of socio-nature processes within waterscapes and to invest in developing further methodologies for interdisciplinary research.

7.9 Epilogue: a critical reflection on the research

Perhaps more than anything, this dissertation represents the scholarly journey I went through during the past eight years, both in terms of expanding my theoretical understanding as well as developing academic skills. So my answer to the question what would I have done differently if I could redo this research would be "*everything*" because I am not longer the one I was before, yet at the same time it would be "*nothing*" because I needed this particular process to reach where I am right now. Nevertheless, there have been a few choices I made and a few challenges that I encountered that influenced the outcomes of this research and, following the reflexive science tradition, I would like to reflect on these key elements that shaped this study.

First, recognizing that knowledge is always partial and situated and detachment can never be fully achieved, I chose to actively engage with the subject of my research and to explicitly state my viewpoint throughout this dissertation. I perceive the world as structurally unjust with unequal chances depending on when and where you are born and in which social category you are placed based on social constructs such as race, gender, age, sexual orientation and class. This understanding of the world has not changed since the start of my research, rather this study has given me more profound insight in how social differences are produced and maintained and how this unfolds within the water domain. Certainly, when you look for inequity, you will find inequity, so that is why I made my assumptions and research methodology transparent to allow other scientists to scrutinize the outcomes of this research and complement or challenge the finding based on other empirical evidence or through a different reasoning. More than ever my encounters with the people I interviewed made me aware of my own social identity as a white young woman. Everywhere I was an outlier, among the small-scale farmers my skin was too light and among the commercial farmers my gender was unalike. Still I believe this social identity gave me access to a wide variety of interviewees as few felt threatened by my presence and with most I could engage in an open and revealing conversation. Only in two cases I politely ended the interview prematurely, one time because an interviewee clearly showed disinterest and one time because I could not bear any longer the racists remarks made. In the writing up of this dissertation I struggled with the use of binary social categories; on the one hand I felt the need to show how the water reform processes interacted with (historical) processes of social differentiation, while on the other hand I did not want to contribute to the reproduction of simplistic and stigmatizing social identities based on one's gender, race, ethnicity and/or class. I did my best to point out the complex and contested nature of social identities, to show the diversity of actors within social categories and to discuss the interrelations and dependencies between various groups of actors within society. Nevertheless, sometimes the findings of this research so obviously showed that particular groups benefited from the water reform process while other groups were marginalized that I could not avoid referring to black and white, women and men, subsistence and commercial farmers.

Second, within the research methodology I made some deliberate choices that shaped the outcomes of this study. This research had an exploratory character, especially at the beginning. Even though from the start I concentrated on understanding negotiations over water between socially uneven positioned actors and even though my objective remained to address equity issues, the context of the ongoing water reform processes only emerged later. Moreover, in the beginning I did not yet have a comprehensive theoretical framework nor an elaborated research strategy, these only evolved while carrying out the research. In addition, this research was carried out on a less than part time basis, which forced me to break the research down in manageable parts and as such I carried out the extended case studies consecutively. Even though these choices allowed me to capitalize on the progressive insight I obtained in each next case study, it also meant the case studies are somewhat fragmented as can been seen from the differences in focus, terminology and theoretical concepts used in the first case study located in Tanzania and the last case study located in Zimbabwe. Nevertheless, in this chapter I have attempted to bring the four cases together in an incorporated comparison to illuminate how the water reforms in the case study countries form part of, and mutually constitute, a larger world-historical process, while simultaneously highlighting the specific value of each case study in answering the overall research question. However, it should be noted that within this incorporated comparison the analysis of how the cases shape and condition each other has been limited and could have been further elaborated. In addition, this research did not have a solid funding base but instead was supported by various projects and funding sources. This forced me to select case studies partially based on the pragmatic consideration of available funding. It also meant that I could not carry out all fieldwork myself but had to engage MSc students to participate in this research as part of their thesis research. Even though involving multiple researchers can aid the objectivity of the research outcomes, it also gave me less control to ensure the quality of the empirical data. Through close supervision and joint analysis of the collected data, I tried to overcome this disadvantage.

Third, a main challenge that I encountered is related to the interdisciplinary nature of this research. Since methodologies for interdisciplinary studies are not readily available, I patched together, like a 'true' bricoleur, different theories into an analytical framework and combined different research approaches that I deemed relevant to do justice to the object of this study. This process has been formed by my educational background, albeit perhaps in a peculiar manner; being originally trained as a river engineer, I first ventured fully into social sciences, intrigued by all that this vast scientific discipline has to offer to a newcomer. While my fieldwork progressed I was confronted with the agency of the physical environment in shaping how water reform processes unfold within waterscapes. This sparked my renewed appreciation of physical processes and the role of infrastructure, which affected my data collection strategy in order to include more technical and/or quantitative data. With hindsight, perhaps this 'reversed' process of bricolage has affected my research and my main argument could have been further strengthened if I would have collected more quantitative data from the start of this research, such as data on water distribution, water use, land use, crop yields and the capacity of hydraulic infrastructures. What I also struggled with was finding the right depth of this research in the sense that my choice to carry out four case studies and aim for an interdisciplinary approach also meant I had less time per case and less time per discipline to collect and analyze data. Of course these two choices within the research approach have been essential; without these I could not have produced the same research findings, though at times it was challenging for me to balance the need for detail in the individual case studies and the aim for studying broader processes. Another challenge was that I placed this research within an emerging school of thought, critical institutionalism. Currently it is still a loose group of scholars originating from various disciplines, each with their own focus and each with different nuances, who might not even identify themselves as critical institutionalist scholars. A common terminology and conceptual understanding still needs to crystallize which sometimes made me wonder if I was dealing with apples and oranges or if they actually did say the same but just using different wording. Moreover, the school of thought is mainly defined in opposition of what is referred to as the mainstream school of thought, neo-institutionalism. Even though it is important to position a new paradigm within the scientific establishment, it is even more important to be able to clearly articulate what the theory entails rather than what it does not stand for. I tried to contribute to this, though this has not been an easy task.

References

Adams, W.M., Potkanski, T. and Sutton, J.E.G. (1994). Indigenous farmer-managed irrigation in Sonjo, Tanzania. *The Geographical Journal* 160(1): 17-32.

Abers, R. N. (2007). Organizing for governance: Building collaboration in Brazilian river basins. *World Development* 35 (8): 1450–1463.

Ahlers, R. (2005). Fixing water to increase its mobility: the neoliberal transformation of a Mexican Irrigation District. PhD dissertation, Ithaca, Cornell University.

Ahlers, R. (2010). Fixing and Nixing: The Politics of Water Privatization. *Review of Radical Political Economics* 42(2): 213-230.

Ahlers, R. and Zwarteveen, M. (2009). The Water Question in Feminism: Water Control and Gender Inequities in a Neo-liberal Era. *Gender, Place and Culture* 16(4): 409-426.

Ahlers, R., Cleaver, F., Rusca, M. and Schwartz, K. (2011). Informal Space in the Urban Waterscape: Disaggregation and Co-Production of Water Services. *Water Alternatives* 7(1): 154-167.

Alden, C. and Anseeuw, W. (2006). Liberalisation versus Anti-imperialism: The impact of narrative on Southern Africa land policies since Zimbabwe's fast-track. Paper presented at the International colloquium on at the frontier of land issues, Montpellier, France.

Allan, T. (1999). Productive efficiency and allocative efficiency: why better water management may not solve the problem. *Agricultural Water Management* 40(1): 71-75.

Anderson, A., Karar, E., Farolfi, S., (2008). Synthesis: IWRM lessons for implementation. *Water SA* 34 (6): IWRM Special Edition.

Andrews, C. (2007). Rationality in Policy Decision Making. In Fischer, F., Miller, G.J., Sidney, M.S. (eds.) *Handbook of Public Policy Analysis, Theory, Politics and Methods*. CRC Press. Taylor & Francis Group: 161-171.

Atkinson, D. and B. Busher (2006). Municipal commonage and implications for land reform: A profile of commonage users in Philippolis. *Agrekon* 45(4).

Atkinson, P. and Hammersley, M. (2007). *Ethnography, Principles in Practice, Third Edition*. New York, Routledge.

Babbie, E. and Mouton, J. (1998). *The Practice of Social Research, South African Edition*. Cape Town, Oxford University Press Southern Africa Limited.

Bachrach, P. and Baratz, M.S. (1962). Two faces of power. *The American Political Science Review* 56: 947-52.

Bakker, K. (2000). Privatizing water, producing scarcity: The Yorkshire drought of 1995. Economic Geography 96(1): 4-27.

Bakker, K. (2002) From state to market?: Water mercantilizacion in Spain. *Environment and Planning A* 34(5): 767-790.

Bakker, K. (2003). A political ecology of water privatization. *Studies in Political Economy* 70: 35-58.

Bakker, K. (2005) Neoliberalizing Nature? Market Environmentalism in Water Supply in England and Wales. *Annals of the Association of American Geographers* 95(3): 542-565.

Bakker, K. (2007). The 'commons' versus the 'commodity': alter-globalization, anti-privatization and the human right to water in the global South. *Antipode* 39(3):430-455.

Barham, E. (2001). Ecological boundaries as community boundaries: the politics of watersheds. *Society and Natural Resources* 14: 181-191.

Bavinck, M. (2005). Understanding fisheries conflicts in the south: A Legal Pluralist Perspective. *Society and Natural Resources* 18: 805-820.

Benería, L. (1999). Globalization, gender and the davos man. *Feminist Economics* 5(3): 61-83.

Benería, L. (2004). *Gender, Development, and Globalisation*. New York, Routledge.

Bentzon, A.W., Hellum, A., Stewart, J., Ncube, W. and Agersnapt, T. (eds.) (1998). *Pursuing Grounded Theory in Law: South-North Experiences in Developing Women's Law*. Mond Book, Harare, Zimbabwe.

Bhatt, Y., Bossio, D., Enfors, E., Gordon, L., Kongo, V. Kosgei, J.R., Makurira, H., Masuki, K., Mul, M.L. and Tumbo, S.D. (2006). Smallholder System Innovations in Integrated Watershed Management (SSI): Strategies of water for food and environmental security in drought-prone tropical and subtropical agro-ecosystems. Colombo, Sri Lanka, *IWMI Working Paper* 109.

Biggs, H. C., Breen, C. M. and Palmer, C. G. (2008). Engaging a window of opportunity: Synchronicity between a regional river conservation initiative and broader water law reform in South Africa. *International Journal of Water Resources Development*, 24(3): 329–343.

Blaikie, P. (1985) The Political Economy of Soil Erosion in Developing Countries. London, Longman Publishing.

Blaikie, P. (2006). 'Is Small Really Beautiful?' Community-Based Natural Resources Management in Malawi and Botswana. *World Development* 34(11):1942-1957.

Blomquist, W. and Schlager, E. (2005). Political pitfalls of integrated watershed management. *Society and Natural Resources* 18: 101-117.

Blumberg, D. and Jacobson, D. (1997) New frontiers: remote sensing in social science research. *The American Sociologist* 28(3): 62-68.

Boelens, R. (2008). The rules of the game and the game of the rules: normalization and resistance in Andean water control. PhD dissertation, Wageningen University, The Netherlands.

Boelens, R., Zwarteveen, M., Roth, D. (2005). Legal complexity in the analysis of water rights and water resources management. In Roth D., Boelens R., Zwarteveen M. (eds.) *Liquid relations: contested water rights and legal complexity*. Rutgers University Press: 1-19.

Bolding, A., (2004). In Hot Water: A Study on Socio-Technical Intervention Models and Practices of Water Use In Smallholder Agriculture, Nyanyadzi, Catchment, Zimbabwe. PhD dissertation, Wageningen, Wageningen University.

Bond, P. (2004). Water Commodification and decommodification Narratives: Pricing and Policy Debates from Johannesburg to Kyoto to Cancun and Back. *Capitalism, Nature and Socialism* 15(1): 9-25.

Bond, P. (2005). *Elite transition: from apartheid to neoliberalism in South Africa*. UKZN Press, http://www.ukzn.ac.za/ccs (accessed on 21 March 2009).

Bond, P. (2006). *Talk left, walk right: South Africa's frustrated global reforms*. UKZN Press, http://www.ukzn.ac.za/ccs (accessed on 21 March 2009).

Bond, P. (2007). Transcending two economies? Renewed debates in South African political economy. Special Issue of AFRICANUS, *Journal of Development Studies*, 37 (2).

Bossio, D., Jewitt, G. and Van der Zaag, P. (2011) Smallholder system innovation for integrated watershed management in Sub-Saharan Africa. *Agricultural Water Management* 98(11): 1683-1686.

Bourblanc, M. (2012). Transforming water resources management in South Africa: catchment management agencies and the ideal of democratic development. *Journal of International Development* 24:637-648.

Bourdieu, P. (1977). *Outline of a Theory of Practice*. Cambridge and New York: Cambridge University Press.

Bratton, M. and Masunungure, E. (2006). Popular reactions to state repression: operation murambatsvina in Zimbabwe. *African Affairs* 106(422): 21-45.

Brewer, G. D. and DeLeon, P. (1983). *The Foundations of Policy Analysis*. Homewood, The Dorsey Press.

Brown, H. I. (1988) *Rationality*. London, Routledge.

Brown, J. (2011). Assuming too much? Participatory water resource governance in South Africa. *The Geographical Journal*, 177 (2): 171–185.

Brown, M. (2006) Survey article: Citizens panels and the concept of representation. *The Journal of Political Philosophy*, 14 (2): 203–225.

Budds, J. (2008). Whose Scarcity? The Hydrosocial Cycle and Changing Waterscapes of La Ligua River Basin, Chile. In Goodman, M.K., Boykoff, M.T., Evered, K.T. (eds.) *Contentious Geographies: Environmental Knowledge, Meaning and Scale*. Aldershot, Ashgate Publishing Limited: 59-78.

Budds, J., and Sultana F. (2013). Exploring Political Ecologies of Water and Development. *Environment and Planning D: Society and Space* 30(2): 275 – 279.

Burawoy, M. (1991). The extended case study method. In Burawoy M., Burton, A., Ferguson, A.A., Fox, K., Gason, J., Gartrell, N., Hurst, L., Kurzman C., Salzinger, L., Schiffman, J., Ui, S. (eds.) Ethnography unbound: power and resistance in the modern metropolis. Berkely, University of California Press: 271-290.

Burawoy, M. (1998). The extended case method. *Sociological Theory*, 16(1):4-33.

Burawoy, M. (2000). Marxism after communism. *Theory and Society* 29:151-174.

Burawoy, M. (2009). *The Extended Case Method: Four Countries, Four Decades, Four Great Transformations, and One Theoretical Tradition*. Berkeley, University of California Press.

Burawoy, M., Blum, J.A., George, S., Gille, Z., Gowan, T., Haney, L., Klawiter, M., Lopez, S.H., O'Riain, S., Thayer, M. (2000). *Global Ethnography: forces, connections and imaginations in a postmodern world*. Berkeley, University of California Press.

Burman, E. (2004). From difference to intersectionality: challenges and resources. *European Journal of Psychotherapy, Counselling and Health* 6:293-308.

Butler, J. (1990). *Gender Trouble: Feminism and the Subversion of Identity*. New York, Routledge.

Castree, N. and Braun, B. (2001). Social Nature: Theory, Practice, Politics. Oxford, Blackwell Publishers.

Castro J.E. (2007). Water Governance in the twentieth-first century. Ambiente & Sociedade 10(2): 97-118.

CETRAD - Centre for Training and Integrated Research in ASAL Development (2010). CETRAD Online Database, www.cetrad.org.

Chikozho, C. (2008). Globaling integrated water resources management: A complicated option in Southern Africa. *Water Resources Management*, 22 (9): 1241–1257.

References

Chikozho, C., and Latham, J. (2005). Shona Customary Practices in the Context of Water Sector Reforms in Zimbabwe, Paper presented at the International workshop on African Water Laws: Plural Legislative Frameworks for Rural Water Management in Africa, January 2005, Johannesburg, South Africa.

Chinguno, N.L. T. (2012) From paper to practice: An analysis of the impact of the water reforms on the water use practices in Nyanyadzi River catchment, Zimbabwe. MSc thesis, UNESCO-IHE, The Netherlands.

CIA - Central Intelligence Agency (2012). The World Fact book. Washington, DC: CIA, www.cia.gov/library/publications/the-world-fact-book.

Cleaver, F. (1999). Paradoxes of Participation: Questioning Participatory Approaches to Development. *Journal of International Development* 11(4): 597-612.

Cleaver, F. (2002) Reinventing institutions: bricolage and the social embeddedness of natural resources management. *The European Journal of Development Research* 14(2): 11-30.

Cleaver, F. (2012) *Development through bricolage: rethinking institutions for natural resource management.* London, Routledge.

Cleaver, F. and Kaare, B. (1998). Social embeddedness and project practice: A gendered analysis of promotion and participation in HESAWA Programme, Tanzania. Centre for Development and Project Planning, Bradford University, United Kingdom.

Cleaver, F. and Elson, D. (1995) Women and Water Resources: Continued Marginalisation and New Policies. The Gatekeeper Series of International Institute for Environment and Development's Sustainable Agriculture Programme 49: 3-16.

Cleaver, F. and Franks, T. (2005). How institutions elude design: river basin management and sustainable livelihoods. Bradford Centre for International Development, University of Bradford, Bradford.

Cleaver, F. and Toner, A. (2006). The evolution of community water governance in Uchira, Tanzania: the implications for equity, sustainability and effectiveness. *Natural Resources Forum* 30 (3): 207-218.

Cleaver, F. and Hamada, K. (2010). 'Good' water governance and gender equity: a troubled relationship. Gender & Development, 18(1): 27 -41.

Cleaver, F. and De Koning, J. (2015). Furthering critical institutionalism. *International Journal of the Commons* 9(1):1-18.

Cocks, M., Dold, A. and Grundy, I. (2002). Challenges facing a community structure to implement CBNRM in the Eastern Cape, South Africa. *African Studies Quarterly* 5 (3).

Conca, K. (2006). *Governing Water: Contentious Transnational Politics and Global Institution Building.* Cambridge: MIT Press.

Cornwall, A. (2003). Whose Voices? Whose choices? Reflections on Gender and Participatory Development. *World Development* 31(8): 1325-1342.

Cornwall A. (2007). Buzzwords and fuzzwords: Deconstructing development discourse. *Development in Practice*, 17 (4–5): 471–484.

Cosgrove, W.J. and Rijsberman, F.R. (2000). *World Water Vision: Making Water Everybody's Vision.* London, Earthscan.

Cousins, B. (2007). Agrarian reform and the two economies: Transforming South Africa's countryside. In: Ntsebeza, L. & Hall, R. (eds.) *The land question in South Africa: The challenge of transformation and redistribution.* HSRC Press, South Africa: 220–245.

Cousins, B. (2007). More than socially embedded: the distinctive character of 'communal tenure' regimes in South Africa and its implications for land policy. *Journal of Agrarian Change* 7: 281–315.

Cousins, B. (2007). A role for common property institutions in land redistribution programmes in South Africa. International Institute for Environment and Development. *Gatekeeper Series* 53.

Coward, J. and Walter, E. (1983). Property in action: Alternatives for irrigation investment. Document prepared for the workshop on Water Management and Policy at the University of Khon Kaen, Thailand.

Cullis, J. and Van Koppen, B. (2009). Applying the Gini coefficient to measure inequality of water use in the Olifants river water management area, South Africa. In: Swatuk L A and Wirkus L (eds.) *Transboundary water governance in Southern Africa: examining unexplored dimensions.* Baden-Baden Nomos: 91-110.

DA - Department of Agriculture South Africa (2004). Broad-based black economic empowerment: Framework for agriculture. DA, Pretoria.

DAFF- Department of Agriculture, Forestry and Fisheries South Africa (2006). Land and Agrarian reform in South Africa: 1994-2006. Paper presented at the international conference on agrarian reform and rural development, Brazil (March 2006).

De Certeau, M. (1984). *The Practice of Everyday Life.* Berkeley, University of California Press.

De Koning, J. (2011). Reshaping Institutions: Bricolage Processes in Smallholder Forestry in the Amazon. PhD dissertation, Wageningen University, The Netherlands.

De Lange, M. (2004). Water policy and law review process in South Africa with a focus on the agricultural sector. In: P. P. Mollinga and A. Bolding (eds.) *The Politics of Irrigation Reform: Contested Policy Formulation and Implementation in Asia, Africa, and Latin America.* London Ashgate: 11-56.

DeLeon, P. (1999). The Stages Approach to the Policy Process: What Has It Done? Where Is It Going? In P. A. Sabatier (ed.) *Theories of the Policy Process*. Oxford, Westview Press: 19-34.

Di Baldassarre, G., Kooy, M., Kemerink, J.S., Brandimarte, L. (2013). Towards understanding the dynamic behaviour of floodplains as human-water systems. *Hydrology and Earth System Sciences Discussion* 10: 3869-3895.

DiMaggio, P. and Powell, W. (1983). The Iron Cage Revisited: Institutional Isomorphism and Collective Rationality in Organizational Fields. *American Sociological Review* 48(2): 147-160.

Dlali, D. M. (2008). Food security in a developmental state. New Agenda, *South African Journal of Social and Economic Policy*, fourth quarter 2008: 43-45.

Douglas, M. (1987). *How Institutions Think*. London, Routledge.

DWA - Department of Water Affairs, South Africa (2000). Guide on the transformation of irrigation boards and certain other boards into water user associations, final document. DWA, Pretoria.

DWAF - Department of Water Affairs and Forestry, South Africa (1997). White Paper on a national water policy for South Africa. DWAF, Pretoria.

DWAF-Department of Water Affairs and Forestry, South Africa (2003). Thukela Water Management Area: overview of water availability and utilization. DWAF (07/000/00/0203), Pretoria.

DWAF - Department of Water Affairs and Forestry, South Africa (2004). Internal strategic perspective: Thukela Water Management Area. Prepared by Tiou and Matji (Pty) Ltd, WRP (Pty) Ltd, and DMM cc on behalf of DWAF, Directorate National Water Resource Planning East. DWAF (P WMA 07/000/00/0304), Pretoria.

DWAF - Department of Water Affairs and Forestry, South Africa (2005). A draft position paper for water allocation reform in South Africa. Towards a framework for water allocation planning. Discussion document. DWAF, Directorate Water Allocations, Pretoria.

Ekers, M. and Loftus, A. (2008) The power of water: developing dialogues between Foucault and Gramsci. *Environment and Planning D: Society and Space* 26: 698-718.

Elson, D. (1995) Gender Awareness in Modeling Structural Adjustment. *World Development* 23(11): 1851-1868

Elson, D. (2012). Review of World Development Report 2012: Gender equality and development. *Global Social Policy* 12(2): 178-183

Enfors, E.I., and Gordon, L.J. (2006). Using water system innovations to escape a poverty trap? Poster presentation WaterNet/WARFSA/GWP-SA, October 2006, Lilongwe, Malawi.

Enfors, E. and Gordon, L., (2007). Analyzing resilience in dryland agro-ecosystems: a case study of the Makanya catchment in Tanzania over the past 50 years. *Land degradation and development* 18: 680-696.

Fairclough N., J. Mulderrig, R. Wodak (2011) Critical Discourse Analysis. Chapter 17 in Van Dijk ed (2011) Discourse Studies: a multidisciplinary introduction. SAGE Publications, London.

Fairclough N. (1996). Discourse Analysis: A Critical View. *Language & Literature* 5(1): 49-56.

Falkenmark, M. and Lundqvist, J. (1999). Towards upstream/downstream hydrosolidarity; introduction. SIWI/IWRA Seminar, Towards upstream/downstream hydrosolidarity, Stockholm: 11-15.

Falkenmark, M., and Folke, C. (2002). The ethics of socio-ecohydrological catchment management: towards hydrosolidarity. *Hydrology and Earth System Sciences* 6(1): 1-9.

Faurès, J.M., Svendsen, M. and Turral, H. (2007). Reinventing irrigation. In: Molden, D. (ed.) Water for Food, Water for Life. Earthscan, London and IWMI, Colombo, Sri Lanka: 353-394.

Fleuret, P. (1985). The social organisation of water control in the Taita hills, Kenya. *American Anthropological Association* 93(1): 91-114.

Fischer, B.M.C. (2008). Spatial variability of dry spells, a spatial and temporal rainfall analysis of the Pangani Basin & Makanya catchment, Tanzania. MSc Thesis, Delft University of Technology, The Netherlands.

Folbre, N. (1994) *Who pays for the kids? Gender and the Structures of Constraint*. New York, Routledge.

Folbre, N. (2012). The Political Economy of Human Capital. *Review of Radical Political Economics Volume* 44(3): 281-292.

Foucault, M. (1979) *Discipline and Punish: The Birth of the Prison*. New York, Vintage.

Foucault, M. (1980). Two Lectures: Selected Interviews and Other Writings 1972-1977. In Gordon, C. (ed.) *Power/Knowledge*. New York, Pantheon:78-165.

Foucault, M. (2000a). Governmentality. In Rabinow, P. (ed.) *Power*. New York, The New Press, 201-222.

Foucault, M. (2000b). The subject and power. In Rabinow, P. (ed.) *Power*. New York, The New Press, 326-348.

Franks, T. and Cleaver, F. (2007). Water Governance and Poverty: A Framework for Analysis. *Progress in Development Studies* 7 (4): 291-306.

Frumkin, P. and Galaskiewicz, J. (2004) Institutional Isomorphism and Public Sector Organizations. *Journal of Public Administration, Research and Theory* 14 (3): 283-307.

Funder, M. and Marani, M. (2015). Local bureaucrats as bricoleurs. The everyday implementation practices of county environment officers in rural Kenya. *International Journal of the Commons* 9(1):87-106.

References

Gibson, K. (2001). Regional subjection and becoming. *Environment and Planning D: Society and Space* 19: 639–667.

Giddens, A. (1984). *The Constitution of Society: Outline of a theory of structuration*. Cambridge, Polity Press.

Goldin, J. A. (2010). Water policy in South Africa: trust and knowledge as obstacles to reform. *Review of Radical Political Economies* 42: 195–212.

Goldin J.A. (2013). The participation paradigm: anathema, praise and confusion. In: Harris, L.M., Goldin, J.A. and Sneddon C. (eds.) *Contemporary Water Governance in the Global South: Scarcity, Marketization and Participation*. London, UK: Routledge: 199-215.

Goldman, M. (2007). How 'Water for all' became hegemonic: the power of the World Bank and its transnational policy networks. *Geoforum* 38:786-800.

GOZ - Government of Zimbabwe (1998a). *Water Act*. Government Printers, Harare.

GOZ - Government of Zimbabwe (1998b). *Zimbabwe National Water Authority Act*. Government Printers, Harare.

Griggs, S. (2007). Rationale Choice in Public Policy: The Theory in Critical Perspective. In Fischer, F., Miller, G.J., Sidney, M.S. (eds.) *Handbook of Public Policy Analysis, Theory, Politics and Methods*. CRC Press, Taylor & Francis Group: 173-185.

Grove, A. (1993). Water use by the Chagga on Kilimanjaro. *African Affairs* 92: 431-448.

Haas, P. (1992) Epistemic communities and international policy coordination. *International Organization* 46(1): 1–36.

Hajer, M. A. (1995). *The politics of environmental discourse: ecological modernization and the policy process*. Oxford, Oxford University Press.

Hall, P.A. and Taylor, R.C.R. (1996). Political science and the three new institutionalisms. *Political Studies* 44(5): 936-957.

Hall, R. (2004). Land and agrarian reform in South Africa: a status report. Programme for Land and Agrarian Studies (PLAAS), School of Government, University of the Western Cape, Cape Town, South Africa.

Hall, S. (1985) Signification, representation, ideology: Althusser and the post-structuralist debates. *Critical Studies in Mass Communication* 2: 91–114.

Haraway, D. (1991). *Simians, Cyborgs and Women: The Reinvention of Nature*. New York: Routledge.

Hardin, G. (1968). The Tragedy of the Commons. *Science, New Series* 162(3859): 1243-1248.

Harris, L.M. (2005). Negotiating inequalities: democracy, gender and the politics of difference in water user groups in south-eastern Turkey. In Adaman, F., and Arsel, M. (eds.) *Turkish environmentalism: between democracy and development*. Aldershot, Ashgate: 185-200.

Harris, L.M. (2009). Gender and emergent water governance: comparative overview of neoliberalized natures and gender dimensions of privatization, devolution and marketization. *Gender, Place & Culture: A Journal of Feminist Geography* 16(4): 387-408.

Hart, G. (2002). *Disabling Globalization: Places of Power in Post-Apartheid South Africa*. Berkeley, University of California Press.

Hart, G. (2006). Denaturalizing dispossession: critical ethnography in the age of resurgent imperialism. *Antipode* 38(5): 977-1004.

Hart, G. (2008). The provocation of neoliberalism: contesting the nation and liberation after apartheid. *Antipode* 40(4): 678-705.

Harvey, D. (1996). *Justice, Nature & the Geography of Difference*. Cambridge & Oxford: Blackwell Publishers.

Harvey, D. (2003). *The New Imperialism*. Oxford, Oxford University Press.

Harvey, D. (2005). *A Brief History of Neoliberalism*. Oxford, Oxford University Press.

Haugaard, M. (2002). *Power: A Reader*. Manchester, Manchester University Press.

Hermans, L.M. and Hellegers, P. (2005). A "new economy" for water for food and ecosystems; synthesis report of E-forum results, FAO/Netherlands International Conference Water for Food and Ecosystems, The Hague.

Hermans, L.M. and Thissen, W.A.H. (2009). Actor analysis methods and their use for public policy analysts. *European Journal of Operational Research* 196(2): 808-818.

Hope, R.A., Jewitt, G.P.W. and Gowing, J.W. (2004). Linking the hydrological cycle and rural livelihoods: a case study in the Luvuvhu catchment, South Africa. *Physics and Chemistry of the Earth* 29: 1209–1217.

Hope, R.A., Gowing, J.W. and Jewitt, G.P.W. (2008). The contested future of irrigation in African rural livelihoods: analysis from a water scarce catchment in South Africa. *Water Policy* 10: 173-192.

Horkheimer, M. (1982). *Critical Theory*. Seabury Press, New York.

Huitema, D., and Meijerink, S. (eds.) (2009). *Water policy entrepreneurs: a research companion to water transitions around the globe*. Cheltenham, Edward Elgar Publishing.

Huitema, D., and Meijerink, S. (2010). Realizing water transitions: the role of policy entrepreneurs in water policy change. *Ecology and Society* 15(2): 26.

Jann, W., & Wegrich, K. (2007). Theories of the Policy Cycle. In Fischer, F., Miller, G.J., Sidney, M.S. (eds.) *Handbook of Public Policy Analysis, Theory, Politics and Methods*. CRC Press, Taylor & Francis Group: 43-62.

Jaspers, F.G.W. (2001). The new water legislation of Zimbabwe and South Africa: Comparison of legal and institutional reform. *International Environmental Agreements: Politics, Law and Economics* 1: 305–325.

Jessop, B. (1998). The rise of governance and the risk of failure: the case of economic development. *International Social Science Journal* 50(155):29-46.

Jessop, B. (2002). *The future of the Capitalist State*. Cambridge, Polity Press.

Johansson, R.C. (2000). Pricing irrigation water: A literature survey. Washington, DC: The World Bank.

Kadirbeyoglu, Z., and Kurtic, E. (2013). Problems and prospects for genuine participation in water governance in Turkey. In: Harris, L.M., Goldin, J.A., Sneddon, C. (eds.). *Contemporary Water Governance in the Global South: Scarcity, Marketization and Participation*. Routledge: 199-215.

K'Akuma, O.A. (2008). Mainstreaming the Participatory Approach in Water Resource Governance: The 2002 water law in Kenya. *Development* 51: 56–62.

Karar, E., Mazibuko G., Gyedu-Ababio, T. and Wetson, D. (2011). Catchment Management Agencies: a case study on institutional reform in South Africa. In: Schreiner, B. & Hassan, R. (eds.), *Transforming water management in South Africa: designing and implementing a new policy framework*. Springer, London: 145–164.

Kemerink, J.S., Ahlers, R., van der Zaag, P. (2007). Driving forces behind the development of water sharing arrangements in traditional irrigation systems in northern Tanzania, Proceedings of 8th WaterNet/WARFSA/GWP-SA symposium, November 2007, Lusaka, Zambia.

Kemerink, J.S., Ahlers, R., van der Zaag P. (2009). Assessment of the potential for hydro-solidarity in plural legal condition of traditional irrigation systems in northern Tanzania. *Physics and Chemistry of the Earth* 34(13-16): 881-889.

Kemerink, J.S., Ahlers, R., van der Zaag P. (2011). Contested water right in post-apartheid South-Africa: the struggle for water at catchment level. *Water SA*, 37(4): 585-594.

Kemerink, J.S., Mbuvi, D., Schwartz, K. (2012). Governance shifts in the water services sector: a case study of the Zambia water services sector. In Katko T., Juuti P.S. and Schwartz K. (eds.) *Water Services Management and Governance: Lessons for a Sustainable Future*. IWA Publishing: 3-11.

Kemerink, J.S., Méndez Barrientos, L.E., Ahlers, R., Wester, P., van der Zaag, P. (2013). Challenging the concept of Water User Associations as the vehicle for transformation: the question of inclusion and representation in rural South Africa. *Water Policy* 15(2): 243-257.

Kemerink, J.S., Munyao, S.N., Schwartz, K., Ahlers, R., van der Zaag , P. (*forthcoming a*) Why infrastructure still matters: unravelling water reform processes in an uneven waterscape in rural Kenya. Under review *International Journal of the Commons*.

Kemerink, J.S., Chinguno, N.L.T., Seyoum, S.D., Ahlers, R., van der Zaag P. (*forthcoming b*) Jumping the water queue: changing waterscapes under water reform processes in rural Zimbabwe. Under review *Natural Resources Forum*.

Kimambo, I.N. (1996). Environmental control and hunger in the mountains and plains of nineteenth-century northeastern Tanzania. In: Maddox G, Giblin J, Kimambo, I.N. (eds.). *Custodians of the land: ecology and culture in the history of Tanzania*. James Currey, London.

Kingdon, J. W. (1984). *Agendas, alternatives and public policies*. New York, Harper Collins.

Kiteme, B. and Gikonyo, J. (2002). Preventing and resolving water use conflicts in the Mount Kenya highland–lowland system through water users' associations. *Mountain Research and Development* 22(4): 332-337.

Komakech, C.H., Mul, M.L., van der Zaag, P. and Rwehumbiza, F.B.R., (2011). Potential of spate irrigation in improving rural livelihoods: case of Makanya spate irrigation system, Tanzania. *Agricultural Water Management* 98 (11): 1719-1726.

Komakech, H.C., van der Zaag, P., Mul, M. L., Mwakalukwa, T.A., Kemerink, J.S. (2012a). Formalization of water allocation systems and impacts on local practices in the Hingilili subcatchment, Tanzania. *International Journal of River Basin Management*, 10(3): 213-227.

Komakech, H.C., van der Zaag, P., van Koppen, B. (2012b). Dynamics between water asymmetry, inequality and heterogeneity sustain canal institutions in Makanya catchment, Tanzania. *Water Policy* 14(5):800-820.

Kongo, V.M. and Jewitt, G.P.W. (2006). Preliminary investigation of catchment hydrology in response to agricultural water use innovations: A case study of the Potshini catchment- South Africa. *Physics and Chemistry of the Earth* 31: 976-987.

Kornov, L. and W. A. H. Thissen (2000). Rationality in decision- and policy-making: Implications for strategic environmental assessment. *Impact Assessment and Project Appraisal* 18(3): 191-200.

References

Kosgei, J.R., Jewitt, G.P.W., Kongo, V.M., Lorentz, S.A. (2007). The influence of tillage on field scale water fluxes and maize yields in semi-arid environments: A case study of Potshini catchment, South Africa. *Physics and Chemistry of the Earth* 32: 1117-1126.

Kirger, N. (2005) ZANU(PF) strategies in general elections, 1980–2000: Discourse and coercion. *African Affairs* 104(414): 1-34.

Kujinga, K. (2002). Decentralizing water management: An analysis of stakeholder participation in the management of water in Odzi sub-catchment area, Save Catchment. *Physics and Chemistry of the Earth* 27: 897-905.

Kujinga, K. and Manzungu, E. (2004). Enduring Contestations: Stakeholder Strategic Action in Water Resource Management in the Save Catchment Area, Eastern Zimbabwe. *Eastern Africa Social Science Research Review* 20(1): 67-91.

Kwezi, L. (2010). The dynamics of struggle for water in post-Apartheid South Africa: Analysis of Negotiations over Water at Catchment Level. MSc thesis, UNESCO-IHE, The Netherlands.

Lahiff, E. (2003). The politics of land reform in southern Africa Sustainable Livelihoods in Southern Africa. Research Paper 19, Institute of Development Studies, Brighton.

Lahiff, E. and Cousins, B. (2005). Smallholder agriculture and land reform in South Africa. *IDS Bulletin* 36 (2): 127-131.

Lankford, B. (2004). Irrigation improvement projects in Tanzania: scale impacts and policy implications. *Water Policy* 6: 89-102.

Latour, B. (1993). *We Have Never Been Modern*. Cambridge, Harvard University Press.

Latour, B. (1996). *Aramis, or the Love of Technology*. London, Harvard University Press.

Laube, W. (2010). Changing the course of history? Contextualising the adoption and implementation of water policies in Ghana and South Africa. In Mollinga P. P., Bhat A., Saravan, V.S. (eds.) *When policy meets reality: Political dynamics and the practice of integration in water resources management reform*. Berlin, LIT Verlag: 61-96.

Laurie, N. (2007) Introduction: How to dialogue for pro-poor water. *Geoforum* 38 (5):753-755.

Leach, M., Mearns, R., Scoones, I. (1999). Environmental Entitlements: Dynamics and Institutions in Community-Based Natural Resource Management. *World Development* 27 (2): 225-247.

Lehman, H.P. (2007). Deepening democracy? Demarcation, traditional authorities and municipal elections in South Africa. *Social Science Journal* 44: 301-317.

Lewis, D. (2009). International Development and the 'Perpetual Present': Anthropological Approaches to the Re-Historicization of Policy. *European Journal of Development Research* 21(1):32-46.

Lewontin, R. and Levins, R. (1997). Organism and environment. *Capitalism Nature Socialism* 8(2): 95–8.

Liebrand, J.W., Zwarteveen, M.Z., Wester, P. and Van Koppen, B. (2012). The deep waters of land reform: land, water and conservation area claims in Limpopo Province, Olifants Basin, South Africa. *Water International* 37(7): 773-787.

Limb, M. and Dwyer, C. (eds.) (2001) *Qualitative methodologies for geographers: Issues and debates*. London, Arnold.

Rindfuss, R.R. and Stern, P.C. (1998) Linking remote sensing and social science: the need and challenges. In: Livermen, D., Moran E.F., Rindfuss, R.R. and Stern, P.C. (eds.) *People and pixels: linking remote sensing and social science*. Washington D.C., National Academy Press:1-27.

Lodge, M. and Wegrich, K. (2005). Control Over Government: Institutional Isomorphism and Governance Dynamics in German Public Administration. *Policy Studies Journal* 33(11): 213-233.

Loftus, A. (2007). Working the socio-natural relations of the urban waterscape in South Africa. *International Journal of Urban and Regional Research* 31: 41-59

Long, N. (1984). *Creating Space for Change: A Perspective on the Sociology of Development*. Wageningen, Wageningen University.

Long, N. (2001). *Development sociology: Actor perspectives*. London and New York , Routledge.

Long, N. and van der Ploeg, J.D. (1989). Demythologizing Planned Interventions: An Actor Perspective. *Sociologia Ruralis XXXIX* 3(4): 226-249.

Long, N. and Long, A. (1992). *Battlefields of knowledge: The interlocking of theory in practice in social research in development*. London, Routledge.

Long, N., and Van der Ploeg, J.D. (1995). Reflections on agency, ordering the future and planning. In: Freks, G. and Den Ouden J. (eds.) *In search of the Middle Ground: Essays on the Sociology of Planned Development*. Wageningen: Wageningen University.

Lowndes, V. (2005). Something old, something new, something borrowed... *Policy Studies* 26(3-4):291-309.

Lundqvist, J., Falkenmark, M. (1999). Introduction Towards upstream/downstream Hydrosolidarity, Proceedings of SIWI/IWRA Seminar Towards upstream/downstream Hydrosolidarity, August 1999, Stockholm, Sweden.

Lyne, M. C. and Darroch, M.A.G. (2004) Land redistribution in KwaZulu-Natal, South Africa: five census surveys of farmland transactions, 1997-2001. BASIS CRSP Management Entity, http://www.basis.wisc.edu (assessed on 28 May 2009).

Macauley, M.K. (2009) *Earth observations in social science research for management of natural resources and the environment*. Washington D.C., Resources for the Future.

Madison, S. D. (2005). Critical ethnography: method, ethics, and performance. Thousand Oaks, CA: Sage.

Magadlela, D. (1999). Whose water Right: A look at irrigators and catchment farmers' water relations in Nyamaropa. In: Manzungu, E., Senzanje, A., and van der Zaag, P. (eds.) *Water for Agriculture in Zimbabwe: Policy and Management Options for the Smallholder Sector*. University of Zimbabwe Publications, Harare: 29-45.

Mail & Guardian (2008). Special Report on Xenophobia in South Africa. http://www.mg.co.za/specialreport/xenophobia (assessed on 20 May 2008).

Mair, L.P. (1969). African marriage and social change. In Phillips, A. (ed.) *Survey of African Marriage and Family Life* (second edition). London, Oxford University Press: 1-172.

Makurira, H., and Mugumo, H. (2003). Water sector reforms in Zimbabwe: the importance of policy and institutional coordination on implementation. Chapter 14 in proceedings of the African Regional workshop on Watershed Management, October 2003, Nairobi.

Makurira, H., Mul, M.L., Vyagusa, N.F., Uhlenbrook, S., Savenije, H.H.G. (2007). Evaluation of Community-Driven Smallholder Irrigation in South Pare Mountains, Tanzania; A case study of Manoo micro-dam. *Physics and Chemistry of the Earth* 32: 1090-1097.

Makurira, H., Mul, M.L., Vyagusa, N.F., Uhlenbrook, S. and Savenije, H.H.G. (2007). Evaluation of community-driven smallholder irrigation in dryland South Pare Mountains, Tanzania: A case study of Manoo micro dam. *Physics and Chemistry of the Earth* 32(15-18): 1090-1097.

Mamdani, M. (1996). *Citizens and subject: contemporary Africa and the legacy of late colonialism*. New Jersey, Princeton University Press.

Manor, J. (2004). User Committees: a potentially damaging second wave of decentralization? *The European Journal of Development Research* 16 (1): 192–213.

Manzungu, E. (2004). Water for All: Improving water resource governance in Southern Africa. International Institute for Environment and Development IIED, London.

Manzungu, E. (2012). The shifting sands of natural resource management in Zimbabwe. Special issue on Natural Resources and Sustainable Development. *Journal of Social Development in Africa* 27: 7-21.

Manzungu, E., and Kujinga, K. (2002). The theory and practice of governance of water resources in Zimbabwe. *Zambezia* XXIX (ii).

Manzungu, E. and Machiridza, R. (2009) Economic-legal ideology and water management in Zimbabwe: implications for smallholder agriculture. *Economics, Management, and Financial Markets* 4(1): 66-102.

Mayoux, L. (1995). Beyond Naivety: Women, Gender Inequality and Participatory Development. *Development and Change* 26(2): 235–258.

Mbeki, M. (2009). *Architects of poverty: why African capitalism needs changing*. Pan MacMillan, South Africa.

McCartney M. and Smakhtin V. (2010). Water storage in an era of climate change: Addressing the challenge of increasing rainfall variability. Colombo, Sri Lanka: IWMI - International Water Management Institute.

McCusker, B. and Ramudzuli, M. (2007). Apartheid spatial engineering and land use change in Mankweng, South Africa: 1963–2001. *The Geographical Journal* 173(1): 56–74.

McGinnis, M.D. (2011). An Introduction to IAD and the Language of the Ostrom Workshop: A Simple Guide to a Complex Framework. *Policy Studies Journal* 39(1):169-183.

McMichael, P. (1990). Incorporating comparison within a world-historical perspective: an alternative comparative method. *American Sociological Review* 55 (3): 385-397.

McMichael, P. (2000) World-systems analysis, globalization, and incorporated comparison. *Journal of world systems research* vi(3) : 68-99.

Meehan, K.M. (2014). Tool-power: Water infrastructure as wellsprings of state power. *Geoforum* 57:215–224.

Mehari, A., Van Koppen, B., McCartney, M. and Lankford, B. (2009). Unchartered innovations? Local reforms of national formal water management in the Mkoji sub-catchment, Tanzania. *Physics and Chemistry of the Earth* 34(4-5):299-308.

Mehari, A., van Steenbergen, F. and Schultz, B. (2005). Water rights and rules, and management in spate irrigation system. International Workshop on African Water Laws: Plural Legislative Frameworks for Rural Water Management in Africa. January 2005, Johannesburg, South Africa.

Meinzen-Dick, R.S. and Pradhan R. (2002). Legal Pluralism and Dynamic Property Rights. CAPRI Working Paper 22. IFPRI: Washington DC.

Meinzen-Dick, R. and Pradhan, R. (2005). Analyzing water rights, multiple uses and intersectoral water transfers. In Roth, D., Boelens, R. and Zwarteveen, M. (eds.) *Liquid relations: contested water rights and legal complexity*. Rutgers University Press: 237-253.

References

Meinzen-Dick, S.R and Nkonya, L. (2007). Understanding legal pluralism in water and land rights: Lesson from Africa and Asia. In van Koppen, B., Giordano, M., Butterworth, J. (eds.) *Community-bases water law and water resources management reform in developing countries*. Wallingford, CAB International: 12-27.

Méndez, L. E. (2010). Devolving resources and power in a context of land and water reform: organising practices, resource transfers and the establishment of a WUA in the Thukela Basin, South Africa. MSc thesis, Wageningen University, The Netherlands.

Méndez, L. E., Kemerink, J.S., Wester, P., Molle, F. *(forthcoming)* The quest for water: Strategizing water control and circumventing reform in rural South Africa. To be submitted to the *International Journal on Water Resource Development*.

Merrey, D.J. (2008). When good intentions are not enough: is water reform a realistic entry point to achieve social and economic equity? In: proceedings of 9th WaterNet/WARFSA/GWP-SA symposium, Johannesburg (October 2008).

Merrey, D. J., Meinzen-Dick, R., Mollinga, P. P. & Karar, E. (2007). Policy and institutional reform: the art of the possible. In: Molden, D. (ed.) *Water for food, water for life: a comprehensive assessment of water management in agriculture*. Earthscan:193–231.

Merrey, D. J., Levite, H. and van Koppen, B. (2009). Are good intentions leading to good outcomes? Continuities in social, economic and hydro-political trajectories in the Olifants river basin, South Africa. In: Molle, F. & Wester, P. (eds.) *River basin trajectories: societies, environment and development*. CAB International, Wallingford: 47–74.

Merry, S.E. (1992). Anthropology, law and transnational processes. *Annual Review of Anthropology* 21: 357-379.

Mills, S. (2003). Power/Knowledge. In: Mills, S. (ed.) *Michel Foucault*. London, Routledge: 67-79.

Mintrom, M. (2000) *Policy entrepreneurs and school choice*. Georgetown University Press, Washington D.C.

Molle, F. (2004). Defining water rights: by prescription or negotiation? *Water Policy* 6 (3): 207–227.

Molle, F. (2008). Nirvana Concepts, Narratives and Policy Models: Insights from the Water Sector. *Water Alternatives* 1(1): 131-156.

Molle, F. and Renwick, M. (2005). The Politics and economics of water resource development: The case of the Walawe river basin, Sri Lanka. IWMI Research Report No 87. Colombo: IWMI.

Molle, F. and Berkoff, J. (eds.) (2007). *Irrigation water pricing: the gap between theory and practice: Comprehensive Assessment of Water Management in Agriculture*. Wallingford, UK: CABI.

Mollinga, P. (2001). Water and politics: levels, rational choice and South Indian canal irrigation. *Futures* 33 (8-9): 733-752.

Mollinga, P. (2003). *On the waterfront: water distribution, technology and agrarian change in a South Indian canal irrigation system*. Orient Longman Private Limited: New Delhi.

Mollinga, P. P. (2008). Water, politics and development: Framing a political sociology of water resources management. *Water Alternatives* 1(1): 7-23.

Mollinga, P.P. (2008). Water Policy - Water Politics: Social Engineering and Strategic Action in Water Sector Reform. In: Scheumann W, Neubert S and Kipping M (eds.) *Water Politics and Development Cooperation*. Berlin: Springer-Verlag Berlin Heidelberg:1-29.

Mollinga, P. P. and Bolding, A. (2004). The politics of irrigation reform: research for strategic action. In: Mollinga, P. P. & Bolding, A. (eds.), The politics of irrigation reform: Contested policy formulation and implementation in Asia, Africa and Latin America. Ashgate, Aldershot. 291–318.

Moore, J. W. (2011). Transcending the metabolic rift: a theory of crises in the capitalist world ecology. *Journal of Peasant Studies* 38(1): 1-46.

Mosse, D. (2004) Is good policy unimplementable? Reflections on the ethnography of aid policy and practice. *Development and Change* 35(4):639–71.

Mosse, D. (2006). Collective Action, Common Property and Social Capital in South India: An Anthropological Commentary. *Economic Development and Cultural Change* 54(3):695-724.

Mosse, D. (2008). Epilogue: The Cultural Politics of Water: A Comparative Perspective. *Journal of Southern African Studies* 34(4): 937-946.

Movik, S. (2011). Allocation discourses: South African water rights reform. *Water Policy* 13 (2): 161–177.

Mshana, R.R. (1992). Insisting upon people's knowledge to resist developmentalism: peasant communities as producers of knowledge for social transformation in Tanzania. PhD Dissertation, Frankfurt University.

Mtisi, S., (2011). Water Reforms During the Crisis and Beyond: Understanding policy and political challenges of reforming the water in Zimbabwe. Working Paper 333, Oversees Development Institute, London.

Mtisi, S., and Nicol, A. (2003). Caught in the Act - New Stakeholders, Decentralization and Water Management Processes in Zimbabwe. Sustainable livelihoods in Southern Africa Project Research Paper No. 14, Institute of Development Studies, Brighton.

Mul, M.L. (2009). Understanding hydrological processes in an ungauged catchment in sub-Saharan Africa. PhD Dissertation, UNESCO-IHE/ TU Delft, Delft.

Mul, M.L., Savenije, H.H.G., Uhlenbrook, S. and Voogt, M.P. (2006). Hydrological assessment of Makanya catchment in South Pare Mountains, semiarid northern Tanzania, Fifth FRIEND World Conference - Climate Variability and Change—Hydrological Impacts-. IAHS, Havana, Cuba: 37–43.

Mul, M.L., Mutiibwa, R.K., Foppen, J.W.A., Uhlenbrook, S. and Savenije, H.H.G. (2007). Mapping groundwater flow systems using geological mapping and hydrochemical spring analysis in South Pare Mountains, Tanzania. *Physics and Chemistry of the Earth* 32: 1015-1022.

Mul, M.L., Mutiibwa, R.K., Uhlenbrook, S., Savenije, H.H.G. (2008a). Hydrograph separation using hydrochemical tracers in the Makanya catchment, Tanzania. *Physics and Chemistry of the Earth* 33: 151-156.

Mul, M.L., Savenije, H.H.G. and Uhlenbrook, S. (2008b). Spatial rainfall variability and runoff response during an extreme event in a semi-arid catchment in the South Pare Mountains, Tanzania. *Hydrological Earth System Sciences Discussion* 13(9):1659.

Mul, M.L., Kemerink, J.S. Vyagusa, N.F., Mshana, M.G., van der Zaag P., Makurira H. (2010). Water allocation practices among smallholder farmers in the South Pare Mountains, Tanzania; can they be up-scaled? *Agricultural Water Management* 98(11): 1752-1760.

Munyao, S. N. (2011). Impact of Formalization of Water Users and Use on Access to Water in Kenya: Case of Likii River Catchment, Upper Ewaso Ngiro North Basin. MSc thesis, UNESCO-IHE, The Netherlands.

Mvungi, A., Mashauri, D. and Madulu, N.F. (2005). Management of water for irrigation agriculture in semi-arid areas: Problems and prospects. *Physics and Chemistry of the Earth* 30(11-16): 809-817.

Nagar, R., (2000). Mujhe Jawab Do! (Answer me!): women's grass-roots activism and social spaces in Chitrakoot (India). *Gender, Place and Culture* 7 (4): 341–362.

NEPAD - New Partnership for African Development (2010). Agriculture in Africa: Transformation and outlook. NEPAD, Johannesburg, South Africa.

Neubert, S., Scheumann, W., van Edig, A. (eds.) (2002) *Reforming institutions for sustainable water management*. Bonn, Deutsches Institut für Entwicklungspolitik.

Nilsson, D. and Nyanchaga, N. N. (2009). East African Water Regimes: The case for Kenya. In: Dellapenna, J.W., Gupta, J. (eds.) *The Evolution of the Law and Politics of Water*. Dordrecht, Netherlands: Springer: 105-120.

Nightingale, A. (2003). A Feminist in the Forest: Situated Knowledges and Mixing Methods in Natural Resource Management. Edinburgh, School of GeoSciences, University of Edinburgh.

Nightingale A. (2011). Bounding difference: Intersectionality and the material production of gender, caste, class and environment in Nepal. *Geoforum* 42: 153-162.

Omer-Cooper, J.D. (1994). *History of Southern Africa* (second edition). Oxford: James Currey Ltd.

O'Reilly, K. (2006). 'Traditional' women, 'modern' water: Linking gender and commodification in Rajasthan, India. *Geoforum* 37: 958-972.

O'Reilly K., Halvorson, S., Sultana, F., Laurie, N. (2009). Introduction: global perspectives on gender-water geographies. Gender, Place and Culture 16(4):381-385.

Ostrom, E. (1990). *Governing the commons: the evolution of institutions for collective action*. New York, Cambridge University Press.

Ostrom, E. (1992). *Crafting Institutions for Self Governing Irrigation Systems*. ICS Press, San Francisco.

Ostrom, E. (1993). Design principles in long-enduring irrigation institutions. *Water Resources Research* 29(7): 1907-1912.

Ostrom, E. (1999). Coping with tragedies of the commons. Annual Review of Political Science 2(1):493-535.

Patrick, M.J., Turton, A.R. and Julien F. (2006). Transboundary Water Resources in Southern Africa: Conflict or cooperation? *Development* 3(49): 22-31.

Peck, J. (2004). Geography and public policy: constructions of neoliberalism. *Progress in Human Geography* 28(3): 392-405.

Peck, J. and Theodore, N. (2010). Mobilizing policy: Models, methods, and mutations. *Geoforum* 41: 169–174

Penzhorn, C. (2005). Participatory research: opportunities and challenges for research with women in South Africa. *Women's Studies International Forum* 28:343-354.

Perret, S.R. (2001) New water policy, irrigation management transfer and smallholding irrigation schemes in South Africa: institutional challenges. FAO, Rome, www.fao.org/landandwater (accessed on 16 August 2009).

Peters, P.E. (2009) Challenges in land tenure and land reform in Africa: anthropological contributions. *World Development* 37(8):1317-1325.

Petty, N. J., Thomson, O.P., Stew, G. (2012). Ready for a paradigm shift? Part 2: Introducing qualitative research methodologies and methods. *Manual Therapy* 17: 378-384.

Pickles, J. and Weiner, D. (1991). Rural and regional restructuring of apartheid: ideology, development policy and the competition for space. *Antipode* 23 (1): 2-32.

References

Pierson, P. (2000). Increasing returns, path dependence, and the study of politics. *American Political Science Review* 94(1): 251-267.

Potkanski, T. and Adams, W.M. (1998). Water scarcity, property regimes and irrigation management in Sonjo, Tanzania. *Journal of Development Studies* 34(4): 86-116.

Power, M. (2000) The Audit Society: Second Thoughts. *International Journal of Auditing* 4: 111-119.

Quinn, N. (2012). Water governance, ecosystems and sustainability: a review of progress in South Africa. *Water International* 37(7): 760-772.

Rajabu, K. R. M. and Mahoo, H.F. (2008). Challenges of optimal implementation of formal water rights systems for irrigation in the Great Ruaha River Catchment in Tanzania. *Agricultural Water Management* 95 (9): 1067-1078.

Ramkolowan, Y. and Stern, M. (2009). The developmental effectiveness of united aid: evaluation of the implementation of the Paris Declaration and of the 2001 DAC recommendation on untying ODS to the LDCS: South Africa country study. Development Network Africa, Pretoria, South Africa.

Rap, E. (2006). The Success of a Policy Model: Irrigation Management Transfer in Mexico. *Journal of Development Studies* 42 (8): 1301- 1324.

Republic of Kenya (1999). Environmental Management and Co-ordination Act. Nairobi, Kenya: Government Printers.

Republic of Kenya (2002). The Water Act. Nairobi, Kenya: Government Printers.

Republic of Kenya (2007). Water Resources Management Rules. Nairobi, Kenya: Government Printers.

Roberts, N.C. and King, P.J. (1991) Policy entrepreneurs: Their activity structure and function in the policy process. *Journal of Public Administration Research and Theory* 1 (2): 147-175.

Robinson, L.W.; Sinclair, J.A. and Spaling, H. (2010). Traditional pastoralist decision-making processes: lessons for reforms to water resources management in Kenya. *Journal of Environmental Planning and Management* 53(7): 847-862.

Roe, E. (1991). Development narratives, or making the best of blueprint development. *World Development* 19(4): 287–300.

Rockström, J., Gordon L., Folke, C., Falkenmark, M. and Engwall, M. (1999). Linkages among water vapour flows, food production, and terrestrial ecosystem services. *Conservation Ecology* 3(2):5Roe, E. (1994). *Narrative Policy Analysis: Theory and Practice*. Durham, Duke University Press.

Ross, R. (1999). *A concise history of South Africa*. Cambridge, Cambridge University Press.

RSA - Republic of South Africa (1996). *Constitution of the Republic of South Africa 1996*. Pretoria, Government Printer, www.info.gov.za/documents/constitution (accessed on 18 July 2008).

RSA - Republic of South Africa (1998a) *National Water Act*. Government Gazette 398, no.19, 182. Cape Town.

RSA - Republic of South Africa (1998b) *Local Government: Municipal Structures Act*. Government Gazette 402, no.19, 614. Cape Town.

RSA - Republic of South Africa (2003) *Traditional Leadership and Governance Framework Amendment Act*. Government Gazette 462, no.25855. Cape Town.

RSA - Republic of South Africa (2004) *Black Economic Empowerment Act*. Government Gazette 463, no.25899. Cape Town.

Rural Focus (2009). Sub-Catchment Management Plan for Likii river Catchment. Nanyuki, Kenya: Water Resources Management Authority. Rhodes, R. A. W. (2006). Policy Network Analysis. In M. Moran, M. Rein and R. E. Goodin (eds.) *The Oxford Handbook of Public Policy*: 423-445.

Rusca, M. and Schwartz, K. (2012). Divergent Sources of Legitimacy: A Case Study of International NGOs in the Water Services Sector in Lilongwe and Maputo. *Journal of Southern African Studies* 38(3): 681-697.

Sachs, J.D., Warner, A., Aslund, A., Fisher, S. (1995) Economic reform and the process of global integration. Washington, Brookings Institution Press.

SAIPRO (2004). Evaluation report of SAIPRO Trust Fund Multi-annual programme. ETC East Africa Ltd, Kenya.

Saletha R. M., and Dinar A. (2005) Water institutional reforms: theory and practice. *Water Policy* 7: 1–19.

Sambu, D. (2011). Water reforms in Kenya: A historical challenge to ensure universal water acess and meet the Millennium Development Goals. PhD Dissertation, University of Oklahoma, USA.

SASSA- South African Social Security Agency (2008) *You and your grants 2009/10*. www.sassa.gov.za (accessed on 24 August 2009).

Savenije, H.H.G. (1999). The role of green water in food production in Sub-Saharan Africa. FAO, Rome, www.fao.org/landandwater (accessed on 21 July 2009).

Savenije, H.H.G. and van der Zaag, P. (2002). Water as an economic good and demand management: paradigms with pitfalls. *Water International* 27(1): 98-104.

Savenije, H.H.G. and van der Zaag, P. (2008). Integrated water resources management: Concepts and issues. *Physics and Chemistry of the Earth* 33(5):290-297.

Schreiner, B. and Hassan, R. (2011). *Transforming water management in South Africa: designing and implementing a new policy framework*. Springer, London.

Scott, J.W. (1986). Gender: A Useful Category of Historical Analysis. *American Historical Review* 91:1053-1075.

Sehring, J. (2009) Path Dependencies and Institutional Bricolage in Post-Soviet Water Governance. *Water Alternatives* 2(1):61-81.

Sheridan, M.J. (2004). The environmental consequences of independence and socialism in North Pare, Tanzania, 1961-88. *Journal of African History* 45: 81-102.

Simpungwe, E. (2006). Water, stakeholders and common ground: challenges for multi-stakeholder platforms in water resource management in South Africa. PhD dissertation, Wageningen University.

Sithole, M. and Makumbe, J. (1997) Elections in Zimbabwe: The ZANU (PF) Hegemony and its Incipient Decline. *African Journal of Political Science* 2(1): 122-139.

Sjaastad, E. and Cousins B. (2008). Formalisation of land rights in the South: an overview. *Land Use Policy* 26: 1-9.

Smaling, A. (1989). Munchausen objectivity: a bootstrap conception of objectivity as a methodological norm. In Bakker, W.J., Heiland, M.E., van Hezewijk, R., Terwee, S. (eds.) *Recent trends in theoretical psychology*. New York, Springer Verlag.

Smith, L. (2004). The Murky Waters of the Second Wave of Neoliberalism: Corporatization as a Service Delivery Model in Cape Town. *Geoforum* 35: 375-393.

Sneddon, C., Harris, L., Dimitrov, R., Ozesmi, U. (2002). Contested Waters: Conflict, Scale, and Sustainability in Aquatic Socioecological Systems. Society and Natural Resources, 15: 663-675.

Sokile, C.S., Kashaigili, J.J. and Kadigi, R.M.J. (2003). Towards an integrated water resource management in Tanzania: the role of appropriate institutional framework in Rufiji Basin. *Physics and Chemistry of the Earth* 28(20-27): 1015-1023.

Sokile, C.S. and van Koppen, B. (2004). Local water rights and local water user entities: the unsung heroines of water resource management in Tanzania. *Physics and Chemistry of the Earth* 29: 1349-1356.

Spiertz, H.L.J. (2000). Water rights and legal pluralism: some basics of a legal anthropological approach. In: Bruns, B.R. and Meinzen-Dick, R.S. (eds.) *Negotiating Water Rights*. London, Intermediate Technology Publications: 245-268.

STATSA - Statistics South Africa (2007). Community survey 2007: Statistical release basic results municipalities. STATSA, Pretoria.

Steyl, I., Versfeld, D. B. and Nelson, P.J. (2000). Strategic environmental assessment for water use: Mhlathuze Catchment, KwaZulu-Natal. Pretoria: Department of Water Affairs and Forestry.

Stiglitz, J.E. (2012). *The price of inequality: How today's divided society endangers our future*. New York, W.W. Norton & Company.

Streeck, W., Thelen, K. (2005). Introduction: Institutional Change in Advanced Political Economies. In: Streeck, W., Thelen, K. (eds.) *Beyond Continuity: Institutional Change in Advanced Political Economies*. Oxford, Oxford University Press: 1-39.

Svubvure , O., Ahlers, R., and Van der Zaag , P. (2011). Representational participation of informal and formal smallholder irrigation in the Zimbabwe water sector: A mirage in the Mzingwane catchment. *African Journal of Agricultural Research* 6(12): 2843-2855.

Swallow, B., Johnson, N., Meinzen-Dick, R. and Knox, A. (2006). The challenges of inclusive cross-scale collective action in watersheds. *Water International* 31(3): 361-376.

Swallow, B.M., Garrity, D.P. and van Noordwijk, M. (2001). The effects of scales, flows and filters on property rights and collective action in watershed management. *Water Policy* 3: 457-474.

Swatuk, L.A. (2005). Challenges to implementing IWRM in Southern Africa. *Physics and Chemistry of the Earth* 30: 872-880.

Swatuk, L. A. (2008). A political economy of water in Southern Africa. *Water Alternatives* 1(1): 24-47.

SWMRG - Soil-Water Management Research Group, Sokoine University of Agriculture, Tanzania (2003). Baseline report: Pangani Basin. In: Smallholder system innovations in integrated watershed management (SSI), first progress report.

Swyngedouw, E. (1997). Power, nature, and the City. The conquest of water and the political ecology of urbanization in Guayaquil, Ecuador: 1880- 1990. *Environment and Planning A*, 29: 311-332.

Swyngedouw, E. (1999). Modernity and hybridity: Nature, regeneracionismo, and the production of the Spanish waterscape, 1890–1930. *Annals of the Association of American Geographers* 89: 443–465.

Swyngedouw, E. (2005) Governance innovation and the citizen: the Janus face of governance-beyond-the-State. *Urban Studies* 42(11):191-2006.

Swyngedouw, E. (2006) Circulations and metabolisms: (Hybrid) Natures and (Cyborg) cities. Special Issue on technonatural time-spaces, *Science as Culture* 15(2): 105-121.

References

Swyngedouw, E. (2009). The political economy and political ecology of the hydro-social cycle. *Journal of Contemporary Water Research and Education* 142: 56–60.

Swyngedouw, E. (2011.) Interrogating post-democratization: Reclaiming egalitarian political spaces. *Political Geography* 30(7): 370-380.

Tickell, A. and Peck, J. (2003). Making global rules: globalization or neoliberalization? In Peck, J. and Yeung, H.W.C. (eds.) *Remaking the global economy: economic-geographical perspectives*. London, Sage: 163-181.

Tilley, S. and Lahiff, E. (2007). Bjatladi Community Restitution Claim. Programme for Land and Agrarian Studies (PLAAS), School of Government, University of the Western Cape.

Tilley, S., Nkazane, N., and Lahiff, E. (2007). Groenfontein-Ramohlakane Community Restitution Claim. Programme for Land and Agrarian Studies (PLAAS), School of Government, University of the Western Cape.

TIP (2004). Farmers' innovations in traditional irrigation improvement. Pipe conveyance system at Kwa Mlombola water user group, Ndambwe village, Mwanga district. TIP, Tanzania.

Thelen, K. (1999). Historical institutionalism in comparative politics. *Annual Review of Political Science 2*: 369-404.

Turpie, J., Ngaga, Y., Karanja, F. (2003). A Preliminary Economic Assessment of Water Resources of the Pangani River Basin, Tanzania: Economic Value, Incentives for Sustainable Use and Mechanisms for Financing Management. Report to IUCN EARO: 1-96.

Turton, A.R. (2002) Towards hydrosolidarity: Moving from resource capture to cooperation and alliances. African Water Issues Research Unit, Pretoria, South Africa.

Turton, A.R., Meissner, R., Mampane, P.M.and Seremo, O. (2004). A hydropolitical history of South Africa's international river basins. University of Pretoria, African Water Research Unit, Water Research Commission (1220/1/04).

Udas, P.B. and Zwarteveen, M. (2005). Prescribing gender equity? The case of the Tukucha Nala Irrigation System, central Nepal. In: Roth, D., Boelens, B. and Zwarteveen, M. (eds.) *Liquid relations: contested water rights and legal complexity*. Rutgers University Press: 21-43.

Uphoff, N., Esman, M. J., Krishna, A. (1998). *Reasons for Success: Learning from Instructive Experiences in Rural Development*. New Delhi, Vistaar Publications.

Uphoff, N., Ramamurthy, P. and Steiner, R. (1999). *Managing irrigation*. Sage Publications, New Delhi.

URT (2004). 2002 Population and Housing Census, Volume IV, District profile, Same, Central Census Office, National Bureau of Statistics, President's Office, Planning and Privatisation, Dar Es Salaam.

Valentine, G., (2007). Theorizing and researching intersectionality: a challenge for feminist geography. *The Professional Geographer* 59 (1), 10–21.

Van der Kooij, S., Zwarteveen, M. and Kuper, M. (2015). The material of the social: the mutual shaping of institutions by irrigation technology and society in Seguia Khrichfa, Morocco. *International Journal of the Commons* 9(1):129-150.

Van der Zaag, P. (1992). *Chicanery at the Canal Changing Practice in Irrigation Management in Western Mexico*. Amsterdam, a CEDLA Publication.

Van der Zaag, P. (2005). Integrated water resources management: Relevant concept or irrelevant buzzword? A Capacity Building and Research Agenda for Southern Africa. *Physics and Chemistry of the Earth* 30(11-16): 867-871.

Van der Zaag, P. (2007). Asymmetry and equity in water resources management; critical governance issues for Southern Africa. *Water Resources Management* 21(12): 1993-2004.

Van der Zaag, P. and Röling, N. (1996). The water acts in the Nyachowa catchment area. In: Manzungu, E. and Van der Zaag, P. (eds.) *The practice of smallholder irrigation: case studies from Zimbabwe*. Harare, University of Zimbabwe Publications: 161-190.

Van der Zaag, P., and Gupta, J. (2008). Scale issues in the governance of water storage projects. *Water Resources Research* 44(10), W10417.

Van der Zaag, P. and Bolding, A. (2009). Water governance in the Pungwe River Basin: institutional limits to the upscaling of hydraulic infrastructure. In: Swatuk, L.A. and Wirkus, L. (eds.) *Transboundary water governance in Southern Africa: examining unexplored dimensions*. Baden-Baden Nomos: 163-177.

Van Dijk, T.A. (2011). *Discourse Studies: A Multidisciplinary Introduction*. London, Sage Publication Limited.

Van Koppen, B., Jha, N. and Merrey, D.J. (2002). Redressing racial inequities through water law in South Africa: revisiting old contradictions? Comprehensive Assessment of Water Management in Agriculture Research Paper 3, IWMI: Colombo.

Van Koppen, B. and Jha, N. (2005). Redressing racial inequities through water law in South Africa, interactions and contest among legal frameworks. In Roth D, Boelens R and Zwarteveen M (eds.) *Liquid relations: contested water rights and legal complexity*. Rutgers University Press, 195-214.

Van Koppen, B. and Schreiner, B. (2014) Moving beyond integrated water resource management: developmental water management in South Africa. *International Journal of Water Resources Development* 30 (3): 543-558.

Van Koppen, B.; van der Zaag, P.; Manzungu, E. and Tapela B. (2014). Roman water law in rural Africa: the unfinished business of colonial dispossession. *Water International* 39(1): 49-62.

Veldwisch, G.J.; Beekman, W. and Bolding, A. (2013). Smallholder irrigators, water rights and investments in agriculture: Three cases from rural Mozambique. *Water Alternatives* 6(1): 125-141.

Vera-Delgado, J. and M. Zwarteveen (2007). The public and private domain of the everyday of politics of water: the construction of gender and water power in the Andes of Peru. *International Feminist Journal of Politics* 9(4): 503-511.

Vijfhuizen, C., (1999). Rain-making, Political conflicts, Gender Images. In: Manzungu, E., Senzanje, A., and van der Zaag, P. (eds.) *Water for Agriculture in Zimbabwe: Policy and Management Options for the Smallholder Sector*. University of Zimbabwe Publications, Harare: 29 - 45.

Von Benda-Beckmann F. (1981). Forum shopping and shopping forums: Dispute processing in a Minangkabau village, *Journal of legal pluralism* 19: 117-159.

Von Benda-Beckmann, F. (1997). Citizens, strangers and indigenous peoples: conceptual politics and legal pluralism. In von Benda-Beckmann, F., von Benda-Beckmann, K. and Hoekema, A. (eds.) *Natural resources, environment and legal pluralism*. Yearbook Law and Anthropology 9: 1-42.

Von Benda-Beckmann, F. (2002). Who is afraid of legal pluralism? *Journal of legal pluralism* 27: 1-46.

Von Benda-Beckman, F. and van Meijl, T. (1999). *Property Rights and Economic Development: Land and Natural resources in Southeast Asia and Oceania*. New York, Columbia University Press.

Von-Benda-Beckmann, F. and Von-Benda-Beckmann, K. (2006). The Dynamic of Change and Continuity in Legal plural Orders. *Journal of Legal Pluralism and Unofficial Law* 53-54: 1-44.

Vyagusa, F.N. (2005). Water allocation practices for smallholder farmers in South Pare Mountains, Tanzania; A case study of Manoo micro dam (ndiva). MSc Thesis, University of Zimbabwe, Harare.

Waalewijn, P., Wester, P. and van Straaten, K. (2005). Transforming river basin management in South Africa: Lessons from the Lower Komati River. *Water International* 30 (2): 184–196.

Warner J., Wester, P. and Bolding, A. (2008). Going with the flow: River basins as the natural units for water management? *Water Policy* 10 (S2): 121–138.

Weaver, O. (1995). Securitization and Desecuritization. In: Lipschutz, R.D. (ed.) *On Security*. Colombia University Press, New York.

Weiss, G. and Wodak, R. (eds.) (2003). *Critical Discourse Analysis: Theory and Interdisciplinarity in Critical Discourse Analysis*. London: Palgrave.

Wester, P.; Merrey, D.J. and de Lange, M. (2003). Boundaries of consent: stakeholders representation in River basin management in Mexico and South Africa. *World Development* 31(5): 797-812.

Wilson, D. (1999). Exploring the Limits of Public Participation in Local Government. *Parliamentary Affairs* 52(2): 246-259.

Wilson, M. and Thompson, L. (eds.) (1969). *The Oxford History of South Africa: Volume1*. London: Clarendon Press.

Wolf, A. (2008). Healing the enlightenment rift: Rationality, spirituality and shared waters. *Journal of International Affairs* 61 (2): 51–73.

Wouters, P. (1999) The relevance and role of water law in the sustainable development of freshwater: From hydro-sovereignty to hydro-solidarity, Proceedings of SIWI/IWRA Seminar Towards upstream/downstream Hydrosolidarity, August 1999, Stockholm, Sweden.

Wright, J.B. (1971). Bushman raiders of the Drakensberg, 1840-1870: a study of their conflict with stock-keeping peoples in Natal. University of Natal Press, Pietermaritzburg.

WWC - World Water Council (2006). Ministerial Declaration of the Fourth World Water Forum, Mexico. Fourth World Water Forum, 21-22 March, Mexico City.

Yin, R. K. (2003) *Case study research: Design and methods*. Third edition. London, Sage Publication.

Zawe, C., (2006). Reforms in Turbulent Times: A Study of the Theory and Practice of Three Irrigation Management Policy Reform Models in Mashonaland, Zimbabwe. PhD Dissertation, Wageningen University, The Netherlands.

Zimmerer, K.S. and Bassett, T.J. (2003) *Political Ecology: An integrated approach to geography and environment-development studies*. New York, The Guilford Press.

Zwarteveen, M. (2006). Wedlock or deadlock? Feminists' attempts to engage irrigation engineers. PhD dissertation, Wageningen University, The Netherlands.

Zwarteveen, M. (2011). Questioning masculinities in water. *Economic and Political Weekly* 46(18): 40-48.

Zwarteveen, M., Roth, D. and Boelens, R. (2005). Water rights and legal pluralism: beyond analysis and recognition. In: Roth, D., Boelens, B. and Zwarteveen, M. (eds.) *Liquid relations: contested water rights and legal complexity*. Rutgers University Press: 254-268.

Annex A: Water allocation practices among smallholder farmers in the South Pare Mountains, Tanzania; can they be up-scaled? [58]

Abstract

The impact of ambitious water sector reforms, that have been implemented in many countries, has not been uniform, especially in Africa. It has been argued that the disconnect between the formal statutory reality at national level and what is happening on the ground may have widened rather than shrunk. There is, therefore, a renewed interest in local water allocation arrangements and how they function. This study looks at water sharing practices and agreements among smallholder farmers in Makanya catchment (300 km^2), which is part of Pangani river basin (42,200 km2) in northern Tanzania. Existing water sharing agreements have been studied in the Vudee sub-catchment (25 km^2), which has about 38 irrigation furrows of which 20 have micro-dams. Five micro-dams are located at the downstream side of the sub-catchment. At the outlet of the Makanya catchment, farmers practice spate irrigation, using the residual flows from the highlands to irrigate. Based on interviews with smallholder farmers and supported by hydrological data water sharing agreements were found to exist among irrigators using the same furrow, among furrows using the same river and at sub-catchment scale. Some agreements date back to the 1940s. They mostly specify water sharing on a rotational basis at all three scales. No water sharing agreements were found at catchment scale, such as between the water users in Vudee sub-catchment and Makanya village. The study concludes that, as a result of the increase in demand for a diminishing resource, tradeoffs between upstream and downstream water uses have emerged at an increasingly larger spatial scale. At the catchment scale, downstream water users have changed their practices to accommodate the changes in the flow. Currently these claims for water do not clash as upstream water users use the base flow (which does not reach downstream anymore) and downstream water users utilise the flood flows. The water sharing arrangements at sub-catchment scale are negotiated through the social networks of the smallholder farmers and are therefore build on the social ties between the communities. However, at catchment scale, the social ties appear relatively weak in addition to the hydrological disconnect; these links are possibly too weak to build new water sharing arrangements on. It may therefore be necessary to involve more formal levels of government, such as Pangani Basin Water Office, to facilitate the negotiation process and create awareness on the inter-linkages of various water uses at catchment scale.

---

[58] This annex is based on Mul, M.L., J.S. Kemerink, N.F. Vyagusa, M.G. Mshana, P. van der Zaag, and H. Makurira (2011) Water allocation practices among smallholder farmers in the South Pare Mountains, Tanzania; can they be up-scaled? *Agricultural Water Management* 98(11): 1752-1760.

A.1 Introduction

Worldwide there is a renewed interest in local water allocation arrangements and how these function. This interest is not only triggered by the steadily increasing demand for water and hence, the growing need for better and more legitimate water allocation decisions at the local level, but also by the comprehensive water sector reforms that have occurred in many countries since the 1990s. Such reforms were often ambitious in scope, taking the national scale as a starting point, with new policies formulated, new laws enacted, new institutions established and new regulations adopted. Yet the impact on the ground has frequently been superficial, especially in Africa (Manzungu, 2004; Sokile et al., 2003; Sokile and van Koppen, 2004; Swatuk, 2005; Van der Zaag, 2005, 2007; Waalewijn et al., 2005; Wester et al., 2003). It has been argued that the disconnect between the formal statutory reality at national level and what happens on the ground may have widened rather than shrunk. This paper considers the local level as a starting point to contribute to a better understanding of why this disconnect exists and of potential ways to bridge it.

Locally developed water allocation arrangements can be surprisingly robust, as indicated by their endurance over time. This has been documented for Eastern Africa where indigenous irrigation development has a long tradition (Adams et al., 1994; Fleuret, 1985; Grove, 1993; Mvungi et al., 2005; Potkanski and Adams, 1998). A better understanding of what it is that makes such arrangements sustainable could provide new ideas of how institutional arrangements at larger scales could be made more effective. Such up-scaling of the principles underpinning local institutional practices could contribute to bridging the identified gap through a bottom-up approach. However, at increasing scales local water allocation arrangements become increasingly sparse, indicating that there are constraints for up-scaling these practices. Spatial scale therefore emerges as an important factor in the analysis of water institutions (see also Barham, 2001; Blomquist and Schlager, 2005; Cleaver and Franks, 2005; Swallow et al., 2001, 2006).

Pangani Basin Water Organisation (PBWO) is managing one of the nine river basins in Tanzania, and is one of the two pilot basins in implementing the water policy (Mehari et al., 2009). A large component of the implementation is issuing water rights to all water users and the development of catchment management institutions (River Basin Organisations), who are responsible for the water allocation at different spatial, from sub-catchment to catchment to river basin level. This paper contributes to understand the driving forces behind the development of local water sharing arrangements by presenting empirical evidence of the water sharing practices among smallholder farmers practising (supplementary) irrigation in indigenous furrow systems in the Makanya catchment, Tanzania. The paper analyzes the links between the water sharing arrangements, irrigation methods and the hydrological regime at different hydrological scales in the catchment and reflects on the impact of spatial scales on the development of water sharing arrangements. The findings presented are based on in-depth semi-structured interviews with over 40 smallholder farmers in the catchment which were carried out in 2005 and 2007. The findings of the interviews were cross-checked through focus group discussions, observations, comparison with existing documentation of the case-study area and by consultations of informants such as extension officers, local authorities and non-governmental organisations (NGO) active in the region. The hydrological data presented in this paper are based on on-site flow measurements and rainfall data collected by the SSI programme between 2004 and 2008 (Bhatt et al., 2006).

This paper first describes the characteristics of the Makanya catchment as well as the furrow systems in the catchment. In the next section, the water allocation practices will be analyzed at four different hydrological scales and linked to the irrigation methods and hydrological (and/or hydraulic) regime of that specific scale. Finally, in the concluding sections the possible reasons for the nonexistence of water sharing arrangements at larger spatial scales are discussed in relation to hydrological regimes and irrigation methods, together with the relevance of the empirical material presented in the paper.

A.2 Furrow systems in Makanya catchment

Makanya catchment (300 km^2) is located in the South Pare Mountains and forms part of the Pangani River basin, Tanzania (42,200 km^2, Figure A.1). It has a bi-modal rainfall pattern, receiving rainfall in two seasons per year: during the period October to January, the season is locally known as Vuli, and between March and May the season is called Masika. Rainfall patterns vary both inter and intra-seasonally and have shown significant trends, in particular the increase in occurrences of dryspells during the Masika season after 1980 (Enfors and Gordon, 2007). In the past, Masika was the main season for crop production. Farmers have shifted to two-season cultivation, in order to secure food production.

In most rainy seasons, there is a need for supplementary irrigation or rainwater harvesting techniques to produce a reliable crop growth, since rainfall in general is not sufficient for cultivating maize, the preferred staple crop (Makurira et al., 2007b). Supplementary irrigation through indigenous furrows has been practiced in this area for more than one hundred years. Some of the furrow systems include a storage structure, which provides extra hydraulic head to supply the furthest downstream farmers in the system (Makurira et al., 2007a). NGO's have been instrumental in enlarging these storage structures and lining canals (TIP, 2004). The catchment can be divided into three areas with different irrigation practices; highlands (1200 m and above), midlands (in the valley, around 900 m) and lowlands (area around Makanya village, 600 m) (Mul, 2009).

In the highlands (rainfall 800 mm a^{-1}), agriculture is practiced throughout the year, with indigenous furrows diverting water from perennial springs for supplementary irrigation during dry-spells in the rainy seasons and full irrigation during the dry season. From a steep escarpment these perennial rivers flow into the midlands (rainfall 600 mm a^{-1}), where the water is used for supplementary irrigation during the rainy seasons. In general, the command area and the capacity of the irrigation system are out of sync. Ideally, during a dry spell, all the farms are irrigated, however, one system in Bangalala shows that less than 10% of the plots are irrigated when it is most needed (Makurira et al., 2007a). Additionally, during these dry spells, runoff from the rivers is low, and even less water is available. Competition for water during these periods is extremely high (Makurira et al., 2007a). The remaining river water, usually the leakages from the diversion structures, continues its flow downhill until it reaches the valley of the catchment where the majority of the runoff recharges the local aquifer under the sandy river bed (Mul et al., 2007). Only flood flows during the two seasons, resulting from high intensity rainfall events and saturated river beds, reach the outlet of the catchment which the lowland farmers divert into their plots for irrigating crops such as maize and beans (rainfall 500 mm a^{-1}) (Komakech et al., 2011). This type of irrigation is known in literature as spate irrigation (see Mehari et al., 2005).

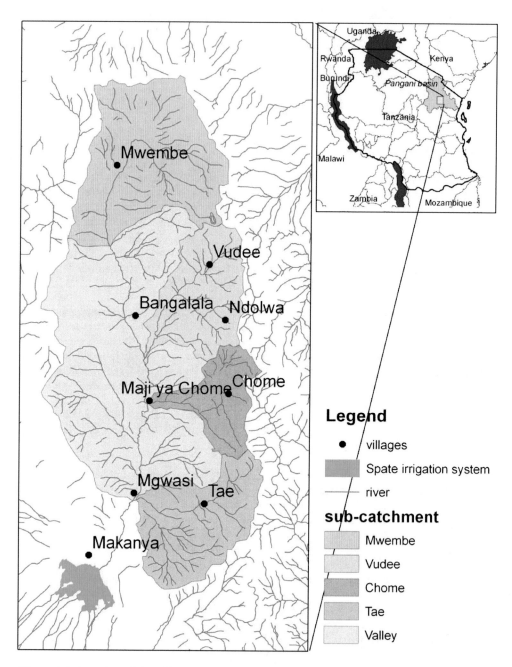

Figure A.1: Location of the different sub-catchment in Makanya catchment within the Pangani Basin in Tanzania.

There are over 100 irrigation furrows in the in Makanya catchment, each supplying water to areas ranging from 0.5 to 400 ha. Most furrow systems are rather rudimentary in terms of materials used. Water is diverted by structures made of rocks, branches and mud. Some aqueducts exist which are made of wooden logs. The furrows are mainly small hand-dug unlined canals and sometimes stretch for several kilometres. Flood flows often destroy the intakes, which need to be rebuilt by the farmers who use the furrow system. It has been reported that the efficiency of these indigenous irrigation schemes are quite low, Turpie et al. (2003) estimated the water losses to be in the order of 80%, which was confirmed by Makurira et al. (2007a), who estimated the total losses of an irrigation system in Makanya catchment to be between 75% and 85%.

Associated with many of the irrigation furrows are micro-dams, located along the irrigation canals locally known as Ndiva (75 have been identified in Makanya, their storage capacity ranges from 200 to 1600 m^3). They are mostly located in the upstream parts of the command area of a furrow system and serve to temporarily store water when nobody irrigates. These are intended to boost the diverted river flow in the furrow when farmers are irrigating. Without such reservoirs, the water would not reach the most distant users because of the large transmission losses (Makurira et al., 2007a). It is important to note that many reservoirs have a fairly long history and were established by local clans before or during the colonial period, and have been given names by the clan members. Over the years most reservoirs have been enlarged to serve the increasing command area. More recently NGOs have assisted irrigators with lining these dams to reduce the losses.

Similar to the Chagga systems described by Grove (1993) the water in the furrow systems is also used for domestic purposes and watering livestock. Especially for households located further away from the communal taps or in areas where the piped water is unreliable in supply, the furrows often serve for multiple uses. The agreements over water sharing concern mainly the furrow systems, as off takes for domestic use are often located upstream of the furrow intakes and are allowed to divert continuously.

A.3 Water allocation practices in Makanya catchment

Water allocation practices among smallholder farmers are found within a relatively small catchment area in Tanzania at four spatial scales, namely: (1) among irrigators sharing one furrow; (2) among furrows along the same river; (3) at sub-catchment scale; and (4) at catchment scale. This section describes the water allocation practices and the underlying institutional arrangements at those spatial scales.

A.3.1 Water allocation among irrigators sharing one furrow

The manner in which water is shared among furrow irrigators is described for the Manoo furrow, located in the outlet of the Vudee sub-catchment in Bangalala village (Figure A1.2). The case-study selected is an irrigation furrow in the midlands, it provides water for supplementary irrigation during the two wet seasons and only full irrigation during the dry season to a very small area of the irrigation system. The furrow takes water from the perennial Vudee river downstream of the confluence of Ndolwa and Upper-Vudee rivers[59] near Bangalala (Figure A.2). It has a total length of approximately 3.5 km. The main canal crosses two significant gullies. Undesired losses to lateral canals and natural drainage systems are minimised by closing off-take points with stones and earth bunds. Not far from the beginning

---

[59] For clarity we differentiate between the part of the Vudee river upstream of the confluence with the Ndolwa river, referred to as Upper-Vudee river, and the part of the river downstream of the confluence referred to as Vudee river.

of the furrow is Manoo micro-dam. Manoo micro-dam is one of the oldest micro-dams in the area which was established in 1936 by the Wadee clan. Over the years the command area and capacity increased. During the Ujamaa villagization policy in the 1970s, people were relocated from the rural areas to villages with basic services. As a result more people joined the irrigation schemes. In 1990, the government of Tanzania abolished natural resources management by the clanship, resulting in Manoo furrow system becoming a property of the whole community. In 2002, the micro-dam was rehabilitated with the assistance of a local NGO, the dam's capacity was increased and the dam was lined to reduce seepage losses and the outlet and diversion structures were modernised. The new capacity of the micro-dam is 1620 m^3 serving about 150 households over an area of 400 ha (Makurira et al., 2007a).

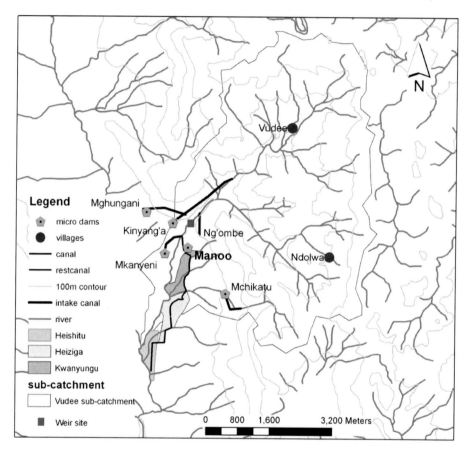

Figure A.2: Location of furrow systems and micro-dams in Bangalala village area.

For management purposes, the Manoo furrow system is divided into three zones, namely Kwanyungu (upstream), Heiziga (midstream) and Heishitu (downstream). The water allocation committee consists of 10 members, including a chairperson, a vice-chairperson, a secretary, a treasurer, the three representatives of the irrigation zones and one additional elderly advisor per irrigation zone. A chairperson, a vice-chairperson, a secretary and a treasurer are elected from the group. The allocation committee meets once a week and decides which zone will receive water at which specific day. Each zone has its own elected representative, locally referred to as Halmashauri, who is responsible for the distribution of water among its members. The farmers are also present at the allocation meeting since other issues are also discussed, such as communal work and conflicts. Directly after the meeting, the farmers can submit their requests for water to the representatives of their zone, who will then allocate the water to the farmers.

A typical allocation turn starts with water diversion from the river at 1600 h into the micro-dam, which fills during the night and once filled the water spills into the furrow nearest to the dam, which is available for whoever is interested. Obviously, the irrigators located in the most upstream zone are at an advantage to use this water. These farmers are often descendants of the Wadee clan. Distributing allocated water starts in the morning and continues until 1600 h and, during this period, diversion into the micro-dam also continues. Distribution of the water to the distribution zones is managed by the representative of the zone. Farmers, receiving water during their irrigation cycle, are responsible for the distribution among themselves and for opening up the bunds. The water is spread on the fields using flood irrigation. Normally up to four beneficiaries receive an allocation per irrigation turn depending on storage available in the micro-dam. This means that if the dam is fully filled, about 400 m3 is allocated to one farmer during his or her irrigation cycle, without considering transmission losses. Makurira et al. (2007a) estimated the transmission losses of the furrow system up to 80% for the most downstream plots.

In allocating the water to the farmers, the representative takes into account whether the farmer already received an allocation during the current season and whether his or her plot and canals are well prepared to receive the water. It is very rare that a farmer gets more than two official allocations within a given season. Some farmers go without an allocation for the entire season, as the capacity of the micro-dam and furrow is far too small to serve the entire command area (Makurira et al., 2007a). Usually there are no requests for irrigation when rainfall is sufficient, but a week or more after a dry spell has started, when crops start experiencing water stress, many requests are received at the same time. However, during such periods river flows may also be low, which means less water available to allocate. When flows are extremely low, farmers located in the upstream zone get priority. The justification given by irrigators is that this is because of the high transmission losses in the system, and that small water releases would not go far into the furrows. Allocating the water nearer to the source is therefore perceived to be more efficient. During those periods, the farmers agree to irrigate smaller plots and downstream farmers are often able to borrow a small plot in the upstream part of the furrow system to grow crops (Kemerink et al., 2009). Conflicts and disputes are discussed in the water allocation committee. In case the water allocation committee cannot solve the conflict the village committee is involved, who also collect fines.

A.3.2 Water allocation among furrows using the same river

The Bangalala furrows are described as an example of how water is being allocated among furrows which divert water from the same river and belonging to the same village. There are six major furrow systems in Bangalala, five with a micro-dam, of which four divert water from the Vudee sub-catchment, namely the furrows known as Mghungani, Kinyang'a, Mkanyeni and Manoo. Mghungani and Kinyang'a divert water from Upper-Vudee river before the confluence with Ndolwa River, while Makanyeni and Manoo divert water from the Vudee river downstream of the confluence with Ndolwa river (Table A.1; Figure A.2). This section describes how the water from Vudee river is shared between the furrows in the Bangalala village. The Upper-Vudee and Vudee rivers are perennial rivers, although during low flows the water is insufficient to supply the entire command area. The adjacent Manoo and Mkanyeni furrow systems divert water at the same location of the Vudee river, each on the other side of the river[60]. There is an agreement on abstraction practices between these furrow systems based on equal access to the river water. Each furrow diverts water for three alternate days in a week from 16.00 hour until 16.00 hour the following day after which the irrigation turn is transferred to the other furrow. Once every two weeks, each furrow system has an irrigation turn on Saturday. This agreement is clearly written in the constitution of Mkanyeni furrow system. The constitution of Manoo furrow system refers to this agreement in the following way: *"Manoo furrow system will co-operate with Mkanyeni furrow system on issues regarding abstraction of water from the river"*. Although no flow measurements are available at the intakes of the furrow systems, the smallholder farmers in both systems express their satisfaction with the current water sharing practices between Mkanyeni and Manoo. Many farmers in the adjacent furrow systems own plots in both systems, which spreads the risk of crop failure and potentially avoid conflicts due to the interdependencies.

Table A.1: Characteristics of the furrow systems located in Bangalala village (Vyagusa, 2005).

Name furrow system	Established (year)	Rehabilitated (year)	Families served	Command area (ha)	Water supply
Manoo	1936	2001	150	400	Vudee
Mkanyeni	1951	2004	70	40	Vudee
Ng'ombe (no micro-dam)	1945	n.a.	40	80	Vudee
Kinyang'a	2000	2004	124	6	Upper-Vudee
Mghungani	1957	2004	115	66	Upper-Vudee
Mchikatu	1959	2000	95	11	Mchikatu

There is no agreement between the Manoo and Mkanyeni furrow systems and the other two upstream furrow systems, Mghungani and Kinyang'a. The upstream furrow systems abstract water every day, even if that means that during dry spells no water is left in the river for the downstream furrows. One of the reasons is that Kinyang'a and Mghungani abstract water from the Upper-Vudee river before the confluence with the Ndolwa. The assumption from Kinyang'a and Mghungani water users is that Manoo and Mkanyeni should use the water that comes from the Ndolwa river, and not the water that comes from the Upper- Vudee river. Therefore, the upstream furrow users do not see the need for an agreement, despite dissatisfaction among the downstream furrow users (Kemerink et al., 2009). However, the major part of the flow in the Vudee river at the intake point of the Manoo and Mkanyeni furrow systems comes from Upper-Vudee and not from Ndolwa, even when Kinyang'a and Mghungani abstract water (Mul et al., 2008a). In other words, even without a mutual agreed water sharing arrangement between the upstream and downstream irrigation systems the hydrological regime at this level in the catchment allows the water to be shared. As all furrow systems are located within one village, conflicts and disputes are solved in the village committee of Bangalala.

---

[60] It should be noted that upstream of the intake of Mkanyeni and Manoo (and upstream of the weir) Ng'ombe furrow system is located. This furrow system does not have a micro dam and it shares in the water abstractions allocated to the Manoo furrow system.

A.3.3 Water allocation at sub-catchment scale

Here, we review the agreements that emerged in the Vudee sub-catchment. At this scale, the agreements are developed between the village administrative units; Vudee, Ndolwa and Bangalala villages in this case. The Vudee sub-catchment drains an area of approximately 25 km^2 at the location of the diversion to Manoo and Mkanyeni furrow systems. In 2002, the sub-catchment had a population of 9,700 growing at a rate of 1.6% per annum (Table A.2; URT, 2004). There are about 38 furrow systems of which 20 have micro-dams. The average size of the micro-dams in Vudee and Ndolwa are smaller (<100 m^3) than the dams in Bangalala (between 400 and 1600 m^3). The water available for Bangalala is affected by water users in the two upstream villages. In the highlands (Ndolwa and Vudee villages) irrigation is mainly used during the dry season and only as supplementary irrigation in the rainy season during dry spells, while in Bangalala supplementary irrigation is almost always needed during the rainy season. Fischer (2008) confirmed this by showing that a critical dry spell in the highlands occurs in is less than 10% of the years, whereas this is almost 90% in the Bangalala area. During the dry season agricultural activities in Bangalala are limited as the flow in the river is inadequate for full irrigation of the entire command area.

Table A.2 Characteristics of villages within the Vudee sub-catchment.

Villages	Vudee	Ndolwa	Bangalala	Total
Population (*source: URT, 2004*)	3,800	2,430	3,470	9,700
Area (km^2)	14.2	8.4	8.8	31.4
Large furrow systems [61](no.)	17	0	1	18
Furrow systems including micro-dam (no.)	6	9	5	20
Irrigated area in rainy season (ha)	200	30	523[62]	753

Agreements between Ndolwa and Bangalala villages

Bangalala has been discussing the issue of water sharing with Ndolwa village since the 1940s. In the first agreement of 1949, Ndolwa agreed to release water for downstream uses. In 1958 this was further defined and one day per week water would be released for downstream users, Ndolwa villagers were not allowed to abstract water from the river for any reason on that day. This agreement was respected by all parties until the 1970s, when Ndolwa began to experience population and economic growth and, an increased pressure on water resources. The 1974 drought saw Ndolwa abandoning the agreement and no longer releasing water for downstream uses. In 1976, the situation returned to normal, but during dry spells, Ndolwa still abstracts water from the river every day, without regard for the consequences for farmers in Bangalala.

At present, people from Ndolwa village claim that flows from the river and streams do not reach Bangalala village during dry spells, even if they do not divert any water. They argue that they have hardly enough water for their own crops during such periods and cannot afford to let water flow downstream. Bangalala villagers are of the opinion that the Mkanyeni and

---

[61] Furrow system without micro-dam owned by more than one family.
[62] Refers to irrigated area in Bangalala during good rainy seasons when water is abundant. During dry spells farmers are only allowed to irrigate part of their lands and during the dry season the irrigated area decreases to almost zero. In higher located Vudee and Ndolwa the irrigated area is almost the same during the rainy season as well as during the dry season, obtaining on average three harvests per year. This explains why the irrigated area per inhabitant in Vudee and Ndolwa is substantial smaller compared to Bangalala.

Manoo furrow systems mainly depend on water from Ndolwa river as the upstream irrigation systems (Mgughani and Kinyang'a) in the village abstract water from Upper-Vudee river. Both villages appreciate the need to find a solution to the water scarcity in the basin, hence, eight water user groups from both villages formed an association of water user groups in 2004 (see Table A.3). This association was formed on advice of an NGO called TIP (Traditional Irrigation Improvement Project; TIP, 2004), which taught farmers in the area to practice soil and water conservation. The association is referred to as 'UNYINDO'[63] and the main focus is "to find new water sources" to be shared by farmers of both villages. Potential new water resources that have been defined are the mountain wetlands, which fall within the administrative boundaries of Ndolwa village, but in the watershed of the villages on the other side of the mountains. However, since the establishment, no formal outcomes have been achieved and the UNYINDO only meets irregularly.

Table A.3: Members of the UNYINDO association of water user groups.

Water user group	Hamlet	Village	No. of families
Heivumba	Masheko	Ndolwa	10
Ndiveni	Ndiveni	Ndolwa	20
Kwanashanja	Mjingo	Ndolwa	40
Kitieni	Kitieni	Ndolwa	40
Kitala	Masheko	Ndolwa	60
Mombo	Mtwana	Ndolwa	60
Mkanyeni	Mkanyeni	Bangalala	70
Manoo	Kwanyungu	Bangalala	150

Agreements between Vudee and Bangalala villages

Bangalala and Vudee villages have an unwritten agreement on the sharing of water based on similar historical water sharing arrangements. Vudee farmers are not allowed to irrigate at night. Abstraction is only allowed for the purpose of filling the micro-dams, but not for run of the river irrigation. At night, water is left to flow for downstream users in Bangalala village where water is diverted to the furrow system, filling the micro-dams, and used for irrigation the next day. On Sundays, Vudee villagers are not allowed to abstract water from the river so that water can be used by the "environment and animals"[64].

Daily variations are observed in the water levels, at the weir site (see Figure A.1 for location). The fluctuations show a clear diurnal pattern, with the highest flows observed early in the morning (Figure A.3), which is consistent with the agreements between Vudee and Bangalala to release water for the downstream village during the night. On Sundays, the decrease of the flow is indeed less than during the other days as indicated in Figure A.3, which corresponds with the intention to allow for environmental flows on Sundays. Another observation is that there is a stronger drawdown on certain week days. This drawdown is attributed to the abstractions from the Ng'ombe furrow system (see Figure A.1) upstream of the weir, which

---

[63] Umoja wa Nyika na Ndolwa (association of water user groups of lowland (Nyika is a sub-village of Bangalala village) and highlands (Ndolwa village)).

[64] It should be noted that in reality this water is used by the downstream Bangalala community, see next Section.

diverts water mainly on Wednesdays within the irrigation turns of members of the Manoo furrow system.

The fluctuations were observed throughout the year, but even more during the dry season as the diversions constitute a substantial amount of the base flow. Hence, it can be concluded that the arrangements agreed between Vudee and Bangalala villages are being adhered to even during periods of low flows. It is also observed that the abstractions are about 10 liter per second, which equates to almost 50% of the total flow during low flows. The fluctuations show the amount of water that is diverted by furrow systems in Vudee village[65,66]. In addition to the abstractions of the farmers in Vudee village adhering to the agreement, potentially there are other abstractions which continuously divert the flows, and therefore reduce the total flow at the weir. During peak flows, the fluctuations at the weir are not observed. This is partly due to the fact that the accuracy of the discharge measurement is reduced with higher flows, and partly because during high flows (in the wet season) the upstream villagers do not need the flow as they do not need to practice supplementary irrigation.

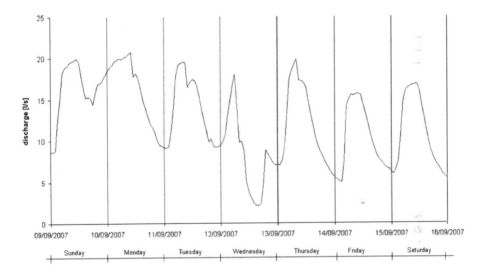

Figure A.3: Typical water level fluctuations at the weir site for one week during dry season as a result of upstream abstractions. Note the shift between abstractions during the day and lowest flow around midnight is due to the lag time of the flow (approximately 1 km h^{-1}).

The development of upstream furrow systems has had a considerable impact on the availability of water in Bangalala. Although the flow at the site of the weir has been perennial as long as people can remember, on some extreme occasions it has been reported to fall dry after a substantial drought period, such as in 1948, 1974, 1997 and in early 2006 (Mul et al., 2006). There does not appear to be a significant increase in the frequency when the river is dry. However, it is reported that the amount of base flow has steadily decreased over the years

---

[65] In this analysis, it is assumed that the furrow systems of Bangalala village upstream of the weir (Kinyang'a and Mghungani) continuously divert water and therefore do not contribute to the fluctuations in the discharge at the weir, however, it does impact the total flow at the weir.

[66] It is assumed that abstractions of the farmers in Ndolwa continuously divert the water as they have no water sharing arrangement with the downstream farmers (see previous Section), this will reduce the flow at the weir but will not influence the fluctuations, the fluctuations are therefore attributed to the agreements with Vudee.

since the 1950s, which is consistent with the increased activities upstream. Bangalala village has repeatedly send representatives to the upstream villages to negotiate agreements. No penalties have been recorded for violating the stated agreements.

A3.4 Water allocation at catchment scale

At the Makanya catchment scale, no water sharing agreements exist between the village of Makanya and upstream villages. As an illustration, we hereby describe the relationship between water users in the Vudee sub-catchment and Makanya village. The most downstream furrow systems of Vudee sub-catchment, Manoo and Mkanyeni, attempt to divert all the water from the river. Upstream of these furrow systems the Vudee river is perennial while downstream the flow becomes intermitted. The only flow downstream is leakage from the diversion structure of the furrow systems, which, within 1 km, completely infiltrates into the sandy river bed. Nowadays, only high intensity rainfall connects the two spatial scales, whereby the flows reach the spate irrigation system in Makanya for several days.

These flood flows, on which spate irrigation relies, have reportedly increased in occurrence and magnitude over the years, with the most extreme event during the El Ni~no year 1997, although in 1951 a heavy flood was also observed (Mul et al., 2006). Moreover, the water source for the spate irrigation system in Makanya is not only limited to the Vudee sub-catchment, but in total four sub-catchments in the South Pare Mountains contribute to generating flood flows to the spate irrigation system (Komakech et al., 2011). Significant spatial variability of the rainfall in the upstream sub-catchments (Mul et al., 2008b) results in a small number of flash floods of sufficient magnitude in Makanya almost each rainy season. During the drought of Vuli 2005 it was observed that, whereas there was complete crop failure in the upstream catchment, crop yields in the spate irrigation system near the intakes were not affected as, despite the drought, an extensive flood lasting up to 10 days and two smaller peak events were received in the lowlands.

A.4 Discussion and conclusion

Water users in the South Pare Mountains have, without the intervention of the national authorities, been able to agree upon sharing available water. As described in this paper water sharing agreements were found to exist among irrigators sharing the same furrow, among furrows using the same river, and at sub-catchment level. Some agreements date back to the 1940s and they mostly specify water sharing on a rotational basis. However, no water sharing agreements were found at a larger spatial scale beyond the sub-catchment scale (25 km^2), such as between the water users in the Vudee sub-catchment and Makanya village.

In the context of the water reform process in Tanzania and the establishment of River Basin Organisations it is interesting to explore why similar arrangements did not evolve over time at catchment scale. One line of reasoning is that in the past, there was no reason for water sharing arrangements at any scale, except within the furrow systems, because no upstream-downstream asymmetries were felt. However, water availability is becoming increasingly scarce, as a result of increasing population density in combination with successful farmer initiatives to harvest more rainfall and to 'drought-proof' or 'climate-proof' their farming systems. At sub-catchment scale, upstream water use directly affected the downstream availability of water (Mul et al., 2006) and the need for water sharing arrangements arose. The social ties between the communities within the Vudee sub-catchment created the opportunity to develop water sharing arrangements. At catchment scale, even more drastic changes occurred, with the perennial river becoming ephemeral. However, at this scale, it is unclear which of the upstream catchments are responsible for the reduction of flow to Makanya, and no clear measures can be taken to assure Makanya a part of the flow. In addition, the social ties are much weaker and therefore no arrangements were developed.

Another line of reasoning is that currently, the upstream furrow irrigators and the downstream spate irrigators do not compete since they use water stemming from different parts of the hydrograph: furrow irrigation upstream typically diverts base flows, whereas the spate irrigation downstream depends on flood flows from four different sub-catchments. In the current setting, this would imply that the increasing water consumption in the upstream parts of the catchment does neither strongly diminish the availability of water downstream nor the frequency of events. Hence, water sharing agreements at the catchment scale are not required. The same line of reasoning holds to a lesser extent for the different furrow systems studied in this paper, as the farmers in the mountains (such as Vudee and Ndolwa) mainly irrigate during the dry season, while the farmers downstream in Bangalala use the water for supplementary irrigation during the rainy season. In this light, it should be noted that furrow systems in the lower reaches developed after the furrow systems in the upper parts of the catchment, which could indicate that the downstream farmers adapted their systems in a way that they would make use of the water which was not already used by the upstream communities. However, due to increased demands for food in the area, the competition over water started to occur, which has eventually triggered the establishment of the agreements between neighbouring villages to share water originating from the same part of the hydrograph. This indicates that similar agreements may also be needed at catchment scale.

Both lines of reasoning indicate that people in the catchment are well aware of the increased competition over water and are actively seeking answers. The water sharing arrangements in Makanya catchment are negotiated through the social networks of the smallholder farmers and are therefore built on the social ties between communities (Kemerink et al., 2009). The hydrological data presented in this paper indicates robustness of locally negotiated water sharing arrangements (see Figure A1.3). However, at the larger spatial scales at which the tradeoffs start to manifest, because the social ties are relatively weak (Cleaver and Franks, 2005): possibly too weak to build new water sharing arrangements on. At levels where no direct competition over water is recognized by the water users and/or at levels where social ties are weak it can be argued that it may be necessary to involve more formal levels of government, such as Pangani Basin Water Office, to assist with the establishment of water allocation agreements. These River Basin Organisations, established as part of the on-going water reform process in Tanzania, could then create awareness on the (indirect) inter-linkages between the various water uses in the basins and facilitate the negotiation processes at larger spatial scales. Potentially mechanisms such as the hydro-solidarity concept (Falkenmark and Lundqvist, 1999; Falkenmark and Folke, 2002; Kemerink et al., 2009) could enhance water sharing practices between the water users at the various levels in the basins.

Both lines of reasoning also raise pertinent questions that beg for answers and that require further research. An argument in favour of maximising water use high up in the catchment, nearest to where it originated as rainfall because of efficiency considerations, raises the question what the impact would be on downstream users, and whether or not downstream users need to be compensated, and if so, how. Reference is made to the recent discussions on the payment for environmental services (Hermans and Hellegers, 2005), but then in reverse mode with payment of compensation flowing from upstream to downstream. The argument in the second line of reasoning that the various furrow systems do not compete over water as they target different parts of the hydrograph needs detailed hydrological and sociological evidence: did the downstream users start using different parts of the hydrograph because of different cropping conditions in their area or was it opportunity driven based on water availability? And what is the precise impact of increased water consumption of many small water users scattered all over the upper parts of the catchment, and how would they impact on the magnitude and frequency of flash floods reaching the catchment's outlet?

The paper has illustrated that the system is complex and dynamic system of bio-physical and social interactions. It showed that hydrological data does not only provide insight in the natural water availability but also provided insights in water allocation practices. At the same time the paper illustrated that institutional arrangements are directly linked to the variable (in time and space) hydrograph. Only a thorough understanding of the social and physical aspects of the water system can lead to a sustainable solution on the longer term as the questions raised above precisely relate to the interaction of the bio-physical and social environment.

Acknowledgements

The work reported here was undertaken as part of the Smallholder System Innovations in Integrated Watershed Management (SSI) Programme funded by the Netherlands Foundation for the Advancement of Tropical Research (WOTRO), the Swedish International Development Cooperation Agency (SIDA), the Netherlands Directorate-General of Development Cooperation (DGIS), the International Water Management Institute (IWMI) and UNESCO-IHE Institute for Water Education. Thanks to WaterNet for supporting the activities of Mr. Vyagusa and Mr. Mshana (MSc fellowships). Implementation on site was assisted by the Soil-Water Management Research Group (SWMRG), Sokoine University of Agriculture, Tanzania.

Annex B: The quest for water: Strategizing water control and circumventing reform in rural South Africa [67]

Abstract

This article shows how large-scale commercial farmers, individually and collectively, are responding to land and water reform processes in the Thukela River Basin, KwaZulu-Natal, South Africa. With a high degree of innovative agency and through four main strategies, commercial farmers have effectively adapted and used their socio-technical systems of water control to neutralise multiple reform efforts that promised to be catalysts for sustainable, inclusive change in the post-apartheid era. Policy by itself will likely continue to fail to facilitate the envisioned transformation if local practices are not sufficiently understood and anticipated by governmental officials, charged with implementation of reform processes.

---

[67] Méndez, L. E., J.S. Kemerink, P. Wester, F. Molle (*forthcoming*) The quest for water: Strategizing water control and circumventing reform in rural South Africa. To be submitted to the *International Journal on Water Resource Development*.

B.1 Introduction

"With water we will wash away the past" were the words of the poet Antjie Krog printed in the White Paper on a National Water Policy (DWAF, 1997). South Africa is acclaimed as one of the most progressive countries in the world when it comes to water policy thinking (Anderson et al. 2008; Biggs et al. 2008; Merrey et al. 2009; Quinn, 2012). The promulgation of a 'revolutionary' water reform process was expressed in a new Constitution that was drafted with active citizen participation after the democratic elections of 1994, driven by constitutional and political imperatives, and later reinforced through legal frameworks: the National Water Services Act and the National Water Act of 1998 (De Lange, 2004).

Although the regulatory frameworks have been in place to support policies for land and water reallocation, implementation of post-apartheid[68] strategies for land and water reform have been slow, confusing and fraught with delays (Cocks et al. 2002; Lahiff, 2003; Hall, 2004; Atkinson and Busher, 2006; Van Koppen and Jha, 2005; Waalewijn et al. 2005; Alden and Anseeuw, 2006; Conca, 2006; Cousins, 2007; Tilley et al. 2007; Tilley and Lahiff, 2007; Merrey et al. 2009; Brown, 2011; Kemerink et al. 2011; Bourblanc, 2012; Quinn, 2012; Kemerink et al. 2013;). Despite the strong emphasis that was placed on correcting the inequities of the past and bridging the gaps between the deprived (mainly black Africans) and the prosperous (mostly white South Africans from European descent), progressive policy has not been supported by progressive implementation (Van Koppen et al. 2002; Alden and Anseeuw, 2006; Conca, 2006; Merrey et al. 2009; Schreiner and Hassan, 2011; Kemerink et al. 2013).

Much attention has been placed on understanding how land reform processes could be successfully implemented to consequently redistribute land (Lahiff, 2003; Hall, 2004; Atkinson and Busher, 2006; Cousins, 2007; Tilley et al. 2007; Tilley and Lahiff, 2007), others have focused on rural institutional transformation and decentralised water resource management that ideally should include former racially disadvantaged groups to actively participate and transform their local realities (Van Koppen et al. 2002; Faysse, 2004; Waalewijn et al. 2005; Brown, 2011; Bourblanc, 2012; Kemerink et al. 2013). However, with a few exceptions (Van Koppen and Jha, 2005; Kemerink et al. 2011; Liebrand et al. 2012), hardly any empirical documentation exists on how local actors, each with their own socio-economic makeup, relations of power and vested interests, are reacting and shaping reform in their favour. Moreover, with the notable exception of Liebrand et al. (2012), little attention has been paid to how white large-scale commercial farmers, either individually or collectively, are circumventing and dealing with land and water reform processes that could directly jeopardize their livelihoods. This paper contributes to fill this knowledge gap.

We suggest that to successfully achieve land and water reform implementation in South Africa, knowledge on the strategic behaviour of all actors should be considered and anticipated for implementation. As noted above, this is especially important since patterns of resource use, access and property have only been partially transformed with the use of formal laws and policies, without critically reflecting on the implementation processes and outcomes of reform to date.

---

[68] Apartheid was a system of legalized racial segregation enforced by the white-dominated national government of South Africa between 1948 and 1994. The government segregated education, medical care, and other public services among different racial groups, and provided black people with inferior services. The educational system was designed to prepare the mass of the black population as labour (Bond, 2006).

In the next section, the theoretical tools are presented. The following section briefly describes the methodology and the local context of the research area as well as, the broader political, social and agrarian context of South Africa. Thereafter, the different water strategies implemented individually by commercial farmers and collectively as Irrigation Boards are presented and analysed. This is followed by a critical reflection on water reform processes and its implementation, emphasizing the implications of local actor's strategies on water reform processes.

B.2 Theoretical considerations

The main focus of this paper is to understand human action amid current water reform processes in South Africa that have the potential, and the promise, to shape the current status quo. In this instance, we analyzed commercial farmers in the case-study catchment. We used the concept of organising practices defined as *"the sets of practices that organise the access to and control over resources such as (land and) water, maintenance machinery, administrative means and other political and economic resources involved in irrigation management"* (Rap, 2004:10) to understand this group's strategic behaviour. We recognize that the commercial farmers are neither a homogenous nor a unified group of actors. However, through time they have developed a strong collective identity based on shared framing of the problems they face and mutual interests they pursue. This collective identity delineates the orientations of their actions and the field of opportunities and constraints in which such actions are to take place (Melucci, 1996; Abers, 2007).

Rap (2004) proposed the analysis of organising practices through two dimensions of control, socio-technical and economic and politico-institutional. The first dimension analyses how different actors mobilise socio-technical networks to control resources. The second dimension views practices, projects and alliances as means to establish economic and politico-institutional control over resources. It builds on the work of Long and van der Ploeg (1995) that view actors' projects as a reflection of specific interests in which resources are mobilised for achieving certain goals and to pattern the social order (Rap, 2004). Resources in this context does not only refer to material resources like land and water or hydraulic infrastructures, but also to resources that serve as means to concretize land and water rights such as labour and money (Mollinga, 2003) as well as social, institutional and political resources that may grant access and enable rights in the first place. As such resources are directly related to relations of power because they are structured properties of social systems and the media through which power is exercised (Giddens, 1984).

The capacity actors have to execute their projects is dependent on their agency. Agency *"refers not to the intentions people have in doing things but to their capacity of doing those things in the first place"* (Giddens, 1984:9). In this it is important not to only understand the capability of actors to take or influence actual decisions, but also the mobilization of 'bias' that is built into institutions through which actors can reinforce certain ideas or decisions over others (Bachrach and Baratz, 1962).

Based on these theoretical considerations we will discuss how the commercial farmer's collective identity was built and analyze the organizing practices they mobilized to manipulate the implementation of water reform process in the case-study catchment in order to pursue their interests.

B.3 Methodology and research context

The paper is based on empirical data collected from April 2010 to August 2010 in the Thukela River Basin in the KwaZulu-Natal province, located in the south-eastern part of South Africa. The actual name and location of the sub-catchment where the research took place will not be revealed due to on-going political sensitivities in the local area, KwaZulu-Natal province and more generally, in South Africa[69]. The findings presented are based on in-depth semi-structured interviews with eighteen commercial farmers, sixteen of which were organised in four Irrigation Boards, and three state officials that worked in the implementation of reform in the study area. The findings of the interviews were triangulated through focus group discussions, observations, comparison with existing literature and by consultations of informants such as local authorities and non-governmental organizations (NGOs) in the area. The sub-catchment has three main tributaries that rise in the Drakensberg Mountains, then flow eastward from a steep escarpment across low mountains of high relief until they reach the lowlands and meet forming one main tributary to the confluence with the Thukela River downstream (*undisclosed reference*).

Despite the sub-catchment's relatively narrow geographical extent, the range of annual net irrigation requirements[70] is between 500 and 1,200 mm per annum (*undisclosed reference*). Irrigation of wheat, soybeans and maize are the main water uses throughout the year, though irrigated fodder crops, mainly sorghum and alfalfa, to support year-round milk production are also large water uses. Relatively high temperatures, extreme rainfall and large irrigation requirements place the sub-catchment under water stress, with water requirements far in excess of the sustainable yield (DWAF, 2004).

The heritage of the colonial and apartheid period and its consequent socioeconomic set up are still clearly visible in the study area. Two major groups live in the sub-catchment. African communities from Zulu descent are located upstream in the sub-catchment, with little agriculture potential or access to water. So-called homelands or Bantustans were once located in these areas during the apartheid regime (Wright, 1971; Kemerink et al. 2011). Herding cattle and small-scale rainfed agriculture are the main activities on relatively small (0.5 to 2 hectares) plots (Kemerink et. al. 2011). In contrast, large-scale farmers from European descent live towards the valley, where the absence of sharp slopes facilitated the development of commercial agriculture. They own ample extensions of land ranging from 30 to 1,500 hectares.

This division of 'white and black land' is the direct result of a process of alienation of indigenous Africans that lasted almost one century between colonial and apartheid rule. During the Afrikaner government (1948-1994) it was aggravated by the implementation of different segregation acts (e.g. Native Land Act (1913), Natives Trust and Land Act (1936), Bantu Authorities Act (1951), Water Act (1956) and the Promotion of Bantu Self-government Act (1959))[71] as well as policies such as the declaration of all Bantu areas as Betterment

---

[69] References used in this paper that directly refer to the case-study area will be indicated with 'undisclosed reference'. For verification purposes the references can be requested from the authors.
[70] Without accounting for conveyance and in-field water losses
[71] The Bantu Authorities Act (1951) provided the legal instrument to establish traditional governance structures (accountable to the white South African government) by appointing local tribal leaders to administer the Bantu areas. In addition, with the Water Act (1956) access to water could only be obtained through riparian rights that accrued to land ownership or through the intervention of the racially discriminatory state. Finally, the Promotion of Bantu Self-government Act (1959) provided the legal framework for separating 'black' spaces from 'white' spaces across the whole nation (Mamdani, 1996; Conca, 2006; McCusker and Ramudzuli, 2007).

Areas[72]. It is estimated that during the apartheid era 83% of agricultural land was in the hands of white commercial farmers (Chikozho, 2008).

Water was also an important instrument of social control. As water rights were under the riparian principle[73] land ownership was crucial for having access to water. In the sub-catchment, large-scale commercial farmers started to organize themselves around water at the start of the twentieth century and Irrigation Boards were legitimised under the former Water Act of 1956. Four Irrigation Boards were formed within the sub-catchment; one along each tributary and the fourth downstream along two irrigation canals. Commercial farmers and Irrigation Boards controlled the majority of water for irrigated agriculture, a total of 6,500 irrigated hectares[74] distributed among 84 members[75]. In addition, control was accelerated and secured by huge investments in dams; one sizeable irrigation dam with a storage capacity of 7.5 million cubic meters was constructed with apartheid government funding partly as a result of strong lobbying of local the Irrigation Boards (*undisclosed reference;* see also Turton et al. *2004)*. The government also funded the construction of two irrigation canals in 1903, when Natal was still a British colony. Recently, the Irrigation Boards have constructed two other dams with a capacity of four and three million cubic meters of storage capacity respectively. On the other hand, the historically discriminated indigenous African farmers had no such organizations and infrastructures and the nature of rain-fed subsistence farming in their communities, still practiced today, strongly limited the potential for improvement and intensification.

With the inclusion of black Africans into the South African nation at the beginning of the 1990s, racially based acts were abolished and new legal frameworks and programs were created to redress spatial, structural and institutional segregation that had caused extreme inequality in land and water distribution (e.g. Communal Land Rights Act[76], National Water Services Act and National Water Act[77] all issued in 1998).

In general, the agrarian reform process has inevitably recreated an environment of tension and anxiety amongst both groups of farmers in the sub-catchment; the small-scale subsistence farmers are impatient and pressing for stronger, faster and effective reforms promised by the

---

[72] Betterment Areas meant the reorganization of existing land use of Bantustans to create Bantu towns and organise land distribution so that each household received a small plot of land in a planned settlement to build a house, plus 1.7-4.2 ha of land for agricultural production and communal areas for pastures. Nevertheless, land was defined as a customary and communal possession and no individual rights to plots were granted. The actual reason behind this spatial engineering was to facilitate increasing population (read: labour force) densities (McCusker and Ramudzuli, 2007). As a result, land tenancy was restricted, the black peasantry eliminated and destined to provide cheap labour for the commercial farming, industry and mining sectors (Mamdani, 1996).

[73] Under the riparian principle the landowners whose properties are adjacent to a body of water have the right to make reasonable use of it. Allocation of water based on the riparian principle makes land ownership important for accessing water. It can be argued that the system of riparian water rights as put forward in the Act resulted in commercial white land-owning farmers having secured access to water.

[74] One irrigated hectare is equivalent to 5,000 m^3/ha per year according to the local definition.

[75] Source from Irrigation Boards water rights and scheduling table for 2010.

[76] The Communal Land Rights Act (CLRA) establishes the transfer of communal land from the State to a community. This entails the devolution of decision-making and the transfer of titles of communal land from the state to a rural community. Under CLRA, tribal authorities (TA) act as 'land administration committees' and make decision on behalf of the communities they represent (Cousins, 2007). Three majors programs recognized by the Constitution have been implemented: Land Restitution, Land Re-distribution and the Reform of Land Tenure.

[77] The National Water Services Act aimed at the definition of provision, extension and management of water services and sanitation, whilst the National Water Act defined water resource management. Water Allocation Reform is one notable policy that aims at reallocating water rights to Historically Disadvantaged Individuals.

government. In juxtaposition, the large-scale commercial farmers are worried about the magnitude of reform that may directly jeopardize their access to resources to which they depend to sustain their commercial enterprises. Furthermore, in order to be competitive with the global market economy, these farmers are pressed to increase their productivity and expand their productions, which might require more land and water resources.

It is within this context that agrarian reform processes are taking place throughout the country. The large-scale commercial farmers continue to be the masters of irrigated agriculture in the study area. They have a long history in being organised around water through one strategic arm: the Irrigation Boards. The Irrigation Boards are trusted organizations that represent their interests at higher political and institutional levels. They have developed a collective identity (see also Abers, 2007) by i) framing their problems in terms of water scarcity and the inability to expand their business and ii) sharing organising practices including sets of strategies to maintain and preferably expand, their access and control over land and water resources to sustain their agribusinesses.

B.4 Water Strategies

The large-scale commercial farmers use four main strategies to protect, control and even expand their access to water resources in the study sub-catchment; 1) the registration of existing lawful water use; 2) creation of 'new water' through the construction of new infrastructures such as dams; 3) the pool system, through the use, distribution and trade of water rights attached to dam reservoirs; and 4) through their tactical leadership in the 'transformation' of Irrigation Boards to inclusive Water User Associations (WUAs).

B.4.1 The registration of lawful water use

The National Water Act (1998) effectively nationalized all water flowing across private or public land. The Department of Water Affairs had to regulate every drop and required licenses from anyone who might use it except for basic human needs (Van Koppen, 2002). Agricultural water users automatically received a new license that fell under the definition of Existing Lawful Use that refers to the water use that lawfully took place in the period two years before the enactment of the 1998 Water Act. In addition, the Department of Water Affairs was tasked with a national compulsory water registration process of water use, including any subsequent redistribution of water of all users throughout the country.

Water registration only started to take place in the year 2000 in the study area. The process seemed simple: Commercial farmers had to register their (historical) water use with the government officials before a certain deadline. In preparation for this process, Irrigation Board meetings were held to explain the registration process to members, analyze the situation and share information (interviews CF6 and CF10). In the case of non-IBs members, registration was more difficult in the sense that they were less prepared, informed and organised to face government officials in comparison to Irrigation Board members (interview CF17). In addition, the leadership of the Irrigation Boards approached Department of Water Affairs and offered their help with the registration process. They lent their offices to government officials so farmers could pass-by the offices to register their water within a determined period of time (interviews CF6 and CF10), avoiding farm visits and possible inspections. Eager to implement reform with limited staff capacity and resources, state officials agreed without realizing the consequences (see also Schreiner and Hassan, 2011).

As shared by commercial farmers and representatives from the four Irrigation Boards, different water registration strategies took place. Some farmers registered less water because they were reluctant to pay for it: *"the more you registered, the more you had to pay"* (interview CF9). They thought that while monitoring was not yet in place and enforced, more water could be taken without any problem and for free. On the contrary, others registered more because they saw it as an opportunity to ensure water for the future (interviews CF4, CF9, CF10; see also Liebrand et al. 2012*).* Since the Department of Water Affairs did not request any formal evidence of historical water use and there were not limits established by produce or land size, commercial farmers registered their historic water rights -in some cases even more- and therefore maintained their riparian rights; *"what happened was that historical riparian water rights were registered under the new act"* (interview CF10). Furthermore, no assessment was made on water availability in the sub-catchment and the future water requirements for potentially emerging farming activities by historically disadvantaged individuals (HDIs)[78] were not taken into account.

Basically, the compulsory government-driven registration processes stated in the 1998 Water Act as a step to redress and reform the water sector did not change anything hitherto. Rather than documenting and registering actual water use, it enhanced and legalized commercial farmers' riparian water rights, strengthening their control of water resources.

B.4.2 Creating 'new' water

A damming and inter-basin transfer culture, supported by multi-million investments in water infrastructure, characterized the water management focus of the apartheid government (Conca, 2006). Unsurprisingly, commercial farmers in the research area are strong advocates of water supply development projects. Most of them privately own at least one irrigation dam located within their properties and if they are members of an Irrigation Board, most have additional access to water stored in the dams owned by the Irrigation Board.

Dams are very important for commercial farmers for two main reasons. First, dams store water for irrigation allowing the development of agriculture all year round. As one commercial farmer summarizes it; *"water is like a Bank if it is stored"* (interview CF13). Second, and perhaps more important, these permanent concrete structures have embodied water rights that are fixed within the structures (see also Mosse, 2008), making water reallocation efforts more difficult. This is no secret to the African National Congress (ANC) government, who discontinued and conditioned water development projects that directly support commercial agriculture owned by these historically advantaged farmers. It is noted that section 27 of the 1998 Water Act specifies that any new supplies of water is reserved as a priority for historically disadvantaged individuals. However, at the time of the fieldwork in 2010, four out of the eighteen interviewed large-scale farmers were in the process of constructing a private dam and/or preparing to get dam permits. Here is how.

In 2005, the Department of Water Affairs introduced the policy of Water Allocation Reform (with the compelling abbreviation WAR) to contribute to Broad Base Black Economic

---

[78] An 'historically disadvantaged individual' is a policy term in the South African context that refers to any person, category of persons or community who is disadvantaged by unfair discrimination before the Constitution of the Republic of South Africa prior to 1993 (Act 200 of 1993) including women and individuals with a disability (DA, 2004).

Empowerment (BBBEE)[79] (DWAF, 2005). In line with WAR, the BBBEE component was introduced as a requirement for all new constructions of water infrastructure. In agriculture, it basically meant that all commercial farmers or Irrigation Boards that wished to build a new dam needed to leave a share of the water stored for the use of historically disadvantaged individuals such as emerging farmers[80].

All of the commercial farmers interviewed disagree with the compulsory BBBEE component. However, they perceive they have no choice but to collaborate and they rather have to 'share' their water than do not have anything at all (interviews CF16 and CF17). As most of historically disadvantaged communities are located upstream farmers and dams, and there are no communities with agricultural land located downstream, the real water that could or would be shared is close to zero. Access to water for irrigation requires designed infrastructure, such as canals and pipes, and management arrangements for water distribution and maintenance. Thus, even if bordering communities decided to engage in irrigated agriculture, it would require i) a considerable financial investment to connect them with new dams, ii) constant management to ensure water gets distributed when its needed, iii) the knowledge to do so and iv) the relationships with commercial farmers who happen to disagree with water reform efforts and are actively engaging in strategies to prevent them from happening.

In the sub-catchment, commercial farmers formed a coalition to build a three million m³ dam of a total cost of 12,8 million South African Rand (ZAR)[81], one year after the passing of the BBBEE law in 2007. With the new dam, some commercial farmers expanded their water rights[82] while no water was set aside for indigenous communities. The commercial farmers proudly describe this achievement as *"the last water license in South Africa without a Black Economic Empowerment component"* (interview CF9).

The construction of the dam illustrates the strength of their agency and the collective socio-technical and economic-political control (Rap, 2004) they actively execute, even at times of land and water reform. It is a reminder that relations of power within social systems enjoy some continuity over time and space (Giddens, 1984).

In order to make the project a reality, farmers had three main hurdles they had to overcome: financial, institutional and bureaucratic. First, there was the *financial* one. Without any governmental support, farmers had to pay for the dam themselves. They needed a sufficient number of interested farmers that could contribute financially to get an affordable loan

---

[79] The premise of BEE was at a first time (in the years to come after the end of apartheid) regarded as simple. It meant the transfer of resources from perceived white-owned business and the South African State to HDIs. The pressure to implement it was both financial and moral. It was thought as a way to amend or rectify the social inequalities of the past, but also it was perceived as a proof of commitment to the 'new era'. According to some critics, most notoriously Moeletsi Mbeki (brother of former president Thabo Mbeki), compliance from business people offered them a chance to keep safeguarding their interests and secure future contracts. But when BEE started to be critiqued as a policy for the empowerment of the few, it was codified in legislation in 2007 as Broad-Based Black Economic Empowerment (BBBEE).

[80] 'Emerging farmer' is a policy term in the South African context that refers to historically disadvantaged individuals who are encouraged and supported by the government to develop their agricultural activities for commercial purposes (DA, 2004). Amongst others, these are individuals and communities who benefitted from land reforms and/or are involved in NGO agricultural projects.

[81] About € 918,624.64 with exchange rate of 1 euro to 13.93 South African Rand (July 10th, 2015)

[82] The amount of water allocated to each commercial farmer depends on their share in the dam based on their financial contribution to the construction of the dam. The sizes of the shares differ considerably, ranging from water to irrigate 1 hectare to water to irrigate 259 hectares. The dam is located on the property of the largest shareholder.

agreement that allowed annual repayment through the collection of annual fees (in addition to the initial financial contribution required). In the end, a total of 41 out of 75 members from three irrigation boards participated.

Second, with a new dam came along a new institutional arrangement. The new infrastructure added complexity to an established water distribution system that enjoyed historic Irrigation Board rules. Since no canals or pipelines were built to distribute the dam's water to its right holders, water had to be distributed through existing irrigation pathways and river tributaries. However, not all farmers had water rights attached to the new dam. This meant that new rules had to be introduced, rearranging property relationships (see also Coward and Walter, 1983; Meinzen-Dick and Pradhan, 2002) and water management realities. In this process, the new institutional set-up benefited those farmers who own rights to the water stored in the new dam, while disabling those without rights (see also Bavinck, 2005). For instance, before the construction of the dam restrictions were applied to share water scarcity among all Irrigation Board members. For this purpose the former by-laws of the Irrigation Boards stated that during the dry season *"water shall be shared per ha of scheduled land owned"*. However, now the rule applies that once the river level drops in the dry winter period, the dam is opened and farmers who hold shares in the dam are allowed to pump their 'concrete' water allocations while the other farmers are not allowed to pump any water.

Finally, getting a dam permit can be a tricky task, especially in times of water reform. Farmers needed a way to navigate bureaucracy and they found their solution through private consultants. With the introduction of the BBEE policy many white bureaucrats exited government and established their consultancy professions. This drained governmental capacity, creating a dependency on consultancy firms to fill the 'know-how' gap (Schreiner and Hassan, 2011). As a result, consultants, who also hold a network of hydrocratic relations, positioned themselves as great mediators to get dam permits for commercial farmers. In the study area, a consultancy firm was hired to deal with the hydrological and hydraulic studies plus to manage the permitting process and negotiate with the Department of Water Affairs. Finally, commercial farmers started construction even if dam permits were not approved. According to commercial farmers, this strategy helped them gain time while the bureaucratic work was being done (interviews CF3, CF4, CF9, CF15, CF16 and CF17). Nevertheless, it is per se a contestation to formal processes and it can be considered as a strategy to force permitting, while it shows the Department of Water Affair's limited monitoring capacity and enforcing authority (interviews O2 and O3; see also Merrey et al. 2009; Schreiner and Hassan, 2011).

B.4.3 The pool system

The pool system acts like an Irrigation Board Water Trust: the members -collective owners of the concrete structure that stores their share in the water- can trade their unused water to other fellow commercial farmers that wish to buy it (interview CF4). These temporary transactions are accorded each year at the beginning of the dry season when farmers have to estimate the amount of water they will use for that period. Water leftovers could be then traded with willing buyers, Irrigation Board members as well as external farmers located downstream of the dams.

This system, in place since the construction of the first dam at the end of the 1980s, allows Irrigation Board farmers to retain property rights to currently unused or newly conserved water, while generating cost savings on water fees from water they would not be using anyway. It also facilitates water transactions, providing temporary water available for other users. As a result, commercial farmers in the study area have two ways of accessing water: one through official water registration with the Department of Water Affairs, and the other informally, through the entrusted water managed internally by the Irrigation Boards.

Even though the pool system was originally designed as an economic strategy, farmers can free themselves from paying annual water fees from unused water, a crucial counter-reform tactic comes out from this organising practice. In order for the Department of Water Affairs to implement any reallocation of water based on equity grounds they first need to call for what is known as a 'compulsory water rights registration' (interviews O2 and O3) as described in previous section. Once water is registered, commercial farmers are obliged to report: i) the total amount of water they are entitled and ii) the actual amount of water they use. According to DWA officials, the leftovers that are without use are theoretically the first water that would be reallocated to other (historically disadvantaged) users because it is assumed that *"this water is lost in the system"* (interview O3). With the pool system however, there are no leftovers as these are temporarily traded each season between commercial farmers. This has given farmers the opportunity to secure their full water rights, protecting their formal allocations against water reallocation efforts.

B.4.4 From Irrigation Boards to 'inclusive' WUA

The promotion of local governance and the transfer of water management to user groups commonly referred to as water user associations (WUAs) has been central in water reform processes in South Africa. According to the 1998 National Water Act each irrigation board would be transformed into a WUA, ideally integrating historically disadvantaged water users. In the sub-catchment, the Department of Water Affairs started transformation with the distribution of an instruction manual that was handled exclusively to commercial farmers as a devoir (see also Liebrand et al., 2012). Instead of fearing the process, commercial farmers saw it as an opportunity to reassert their control and became actively engaged, steering the whole process.

First, they successfully negotiated the boundaries of WUAs with the Department of Water Affairs so that instead of having four separate WUAs, one for each Irrigation Board as mandated by the Water Act, they could have one that would manage the whole watershed. This strengthened their water control over the whole catchment, which now included upstream communities (see also Liebrand et al., 2012). As put by a commercial farmer: *"this is very advantageous for us because now with their (HDIs) integration we will be able to control the whole catchment and protect existing users"* (interview CF10).

Then, focus shifted to the establishment of the WUA membership and leadership. An unbalanced steering committee was elected constituted by four commercial farmers, the four chairmen of the Irrigation Boards, two representatives of the emerging farming sector filled in by a female resident of an urban settlement and a farm worker, neither of whom are emerging farmers, and four associate members of which two have no voting rights[83].

---

[83] For a detailed discussion of this process and considerations on inclusion and representation in the catchment see Kemerink et al. 2013.

The unbalanced platform reflects problems with both inclusion and representation (Kemerink et al. 2013) and instrumentality rather than empowerment is observed (see also Cleaver and Kaare, 1998). As a result, the WUA is a successful negotiating platform for future use, but mainly for established commercial farmers.

B.5 Conclusions

This article shows how large-scale commercial farmers, individually and collectively, are responding to land and water reform processes, in most instances by circumventing them. The socio-technical systems of water control that have historically been managed by these farmers have been effectively adapted and used to contest the ongoing reform processes. With a high degree of innovative agency, they have been able to neutralise the multiple reform efforts that promised to be catalysts for sustainable inclusive change in the post-apartheid era. Based on this case study we argue that the progressive public policies alone have not been capable of facilitating the envisioned transformation. Local socio-technical relations, to a large extent, have determined the outcomes of reform processes and as a result, little change has been experienced in the study area. If local practices are not sufficiently understood and anticipated by government officials, charged with design and implementation of the reform processes, reform will likely continue to fail.

Socio-technical relations play an important role in creating and recreating the contexts in which organising practices are taking place in response to reform processes. For instance, the implementation of the new water law, through water registration and the transformation of Irrigation Boards to WUAs, has failed to produce its objectives when confronted with an unlevel playing field, historically dominated by white commercial farmers, and largely left unchallenged by the government (see also Liebrand et al. 2012).

As documented in this article dams are key resources that are being used to protect and even expand commercial farmers' water rights. Despite BBEE, with the successful construction of dams and with the use of the pool system, farmers have been able to fix their rights in the structures and trade water rights between themselves, leaving no water unused in the system that could be reallocated. Thus, the correlation between agency and water control is directly related to access to this type of infrastructure and may shift as water infrastructure is developed and/or actually shared with historically disadvantaged groups in society.

Acknowledgements

The work presented in this annex was funded by Agris Mundus. Implementation on site was assisted by the School of Bioresources Engineering and Environmental Hydrology and the Centre for Environment, Agriculture and Development of the University of KwaZulu-Natal, South Africa. We thank the respondents in the study catchment for sharing their knowledge and opinions. Our gratitude goes to Prof. Graham Jewitt and Mr. Michael Malinga for facilitating the research.

Samenvatting

Onder druk van internationale organisaties als de Wereldbank hebben veel overheden sinds de jaren 1980 een nieuw beleid gevoerd op het gebied van waterbeheer. De hervorming bestond eruit dat de nadruk werd verlegd van investeringen in de aanleg, exploitatie en onderhoud van waterinfrastructuur, naar de ontwikkeling van regelgeving en het definiëren van principes op grond waarvan water wordt verdeeld onder gebruikers.

Deze gewijzigde aanpak is bekritiseerd op basis van empirisch onderzoek. Het ene onderzoek wijst uit dat met beleidsinterventies vaak nauwelijks iets bereikt wordt. Er wordt beweerd dat dit komt doordat beleid gaandeweg wordt geïnterpreteerd, heronderhandeld en aangepast door actoren actief op verschillende niveaus en met ongelijke machtsposities, hetgeen leidt tot onbedoelde, onvoorspelbare en onvolledige uitkomsten. Het andere onderzoek beweert dat deze nieuwe beleidsaanpak juist wel veranderingen teweeg brengt en zelfs leidt tot dezelfde uitwerkingen in verschillende landen, namelijk een toename van ongelijkheid in de toegang tot, controle over en gebruik van water in stroomgebieden. Dit proefschrift probeert deze wetenschappelijke paradox te ontrafelen door te analyseren in hoeverre, op welke wijze en waarom het nieuwe beleid de toegang tot water in geselecteerde stroomgebieden heeft beïnvloed. Hiervoor is gekeken in welke mate beleidsinterventies van de overheid in de water sector de bestaande institutionele kaders in de maatschappij, die bepalend zijn voor de toegang tot en controle over water binnen de agrarische sector, hebben veranderd. Het doel van dit interdisciplinaire onderzoek is om inzicht te krijgen in hoe de wisselwerking tussen overheidsbeleid en bestaande institutionele kaders stroomgebieden beïnvloedt die historisch zijn gevormd door zowel natuurlijke als sociale processen. Hiertoe worden casussen bestudeerd in vier Afrikaanse landen die sinds de jaren 1980 hun waterbeleid hebben hervormd. Dit zijn Kenia, Zuid-Afrika, Tanzania en Zimbabwe. De hervormingen in deze landen zijn ingegeven door de eerdergenoemde wereldwijde veranderde visie op waterbeheer en hebben als zodanig vergelijkbare uitgangspunten en doelen.

Dit onderzoek bouwt voort op *critical institutionalism* (o.a. Cleaver, 2002; 2012; Cleaver and De Koning, 2015), een theorie waarin een institutioneel kader wordt beschouwd als een uitkomst van dynamische sociale processen dat het menselijk gedrag vormt, reguleert en reproduceert. Deze theorie helpt te verklaren hoe institutionele veranderingsprocessen kunnen leiden tot ongelijke uitkomsten voor verschillende groeperingen in de maatschappij. Om het hedendaagse beleidsvormingsproces te begrijpen, is in dit onderzoek gekozen voor een politiek-theoretisch perspectief waarin beleid wordt gezien als een uitkomst van het bewust handelen van netwerken van beleidsmakers die proberen problemen en ideeën te sturen om zo bepaalde beleidsmodellen te verspreiden die aansluiten bij hun specifieke gedachtegoed en belangen (o.a. Conca 2006, Rap, 2006; molle, 2008; Peck en Theodore, 2010). Ook maakt dit onderzoek gebruik van het concept *waterscape*, waarin de relatie tussen sociale en natuurlijke processen wordt beschouwd als bepalend voor de vorming en herordening van de fysieke kenmerken van een stroomgebied (o.a. Swyngedouw, 1999 Budds, 2008; Mosse, 2008). Met dit concept kan worden geanalyseerd hoe de wisselwerking tussen de bestaande institutionele kaders en de beleidsinterventies de historisch ontstane landschappen veranderen alsmede hoe het hervormingsproces wordt beïnvloed door fysieke processen en objecten zoals infrastructuur voor waterbeheer.

Om de hervormingsprocessen in de watersector te analyseren in de vier stroomgebieden is als onderzoeksmethode gekozen voor de *extended case study method* (o.a. Burawoy, 1991; 1998).

Hiervoor zijn 175 diepgaande interviews gehouden met actoren in de stroomgebieden, waaronder veel zowel grootschalige als kleinschalige boeren. Ook zijn groepsdiscussies en informele gesprekken gevoerd, veldwaarnemingen gedaan en vergaderingen bijgewoond. Beleidsdocumenten, kaarten, satellietbeelden, databanken, wetenschappelijke publicaties en project rapportages zijn bestudeerd binnen dit onderzoek. Elk van de vier casussen richt zich op andere facetten van het hervormingsproces om zo de werking en de gevolgen van de veranderde beleidsaanpak sinds de jaren 80 zo goed mogelijk te begrijpen.

De casus in Tanzania gaat over de onderhandelingen over de toegang tot water tussen boeren in traditionele kleinschalige irrigatiesystemen in de periode dat de eerste stappen werden gezet om het nieuwe beleid in het stroomgebied te introduceren. Deze casus laat het gemêleerde en dynamische karakter van de institutionele kaders zien waarbinnen waterbronnen worden beheerd en gebruikt en hoe de machtsverhoudingen tussen de boeren veranderen door de invoering van het overheidsbeleid. Ook zien we hoe watergebruikers verschillende waarden en normen hanteren in de onderhandelingen over de toegang tot en controle over water, afhankelijk van hun belangen en positie, en hoe het complexe en veelvormige institutionele kader de verdeling van het water tussen de boeren beïnvloedt.

De casus in Zuid-Afrika laat zien dat de hervormingsprocessen in de watersector omstreden zijn en wat dit betekent voor de interacties tussen grootschalige en kleinschalige boeren in het stroomgebied. We zien dat de internationaal geprezen Zuid Afrikaanse waterwetgeving is gebaseerd op verschillende, deels tegenstrijdige, normatieve uitgangspunten. We zien ook dat dit leidt tot een slechts gedeeltelijke uitvoering van de wet binnen de nog steeds sterk gesegregeerde samenleving. Voor dit stroomgebied is geanalyseerd hoe het schijnbaar progressieve beleidsmodel voor de decentralisatie van verantwoordelijkheden naar nieuw opgerichte verenigingen van watergebruikers leidt tot het versterken van de structurele ongelijkheid in termen van toegang tot en controle over water tussen de blanke en zwarte boeren.

De casus in Kenia analyseert de beweegredenen die ten grondslag liggen aan het hervormingsproces in de watersector. Het laat zien dat de beweegredenen niet aansluiten op de behoeften van verschillende groepen watergebruikers. We zien dat slechts een paar historisch bevoorrechte en commercieel georiënteerde boeren hebben geprofiteerd van de nieuwe wetgeving, hetzij door zich aan te passen, hetzij door de effecten ervan te omzeilen. Er komen een aantal onverwachte en nadelige uitkomsten van het hervormingsproces aan het licht voor kleinschalige boeren die lid zijn van verenigingen van watergebruikers. De institutionele diversiteit waarmee deze boeren te maken hebben en het type infrastructuur voor waterbeheer waar zij toegang toe hebben, beïnvloeden die uitkomsten in grote mate.

De laatste casus in Zimbabwe beschrijft hoe de veranderingen in het waterbeleid uitpakken in een stroomgebied dat sterk beïnvloed wordt door een ineenstortende nationale economie. Dit laatste is veroorzaakt door een snelle wijziging in het overheidsbeleid met betrekking tot landhervormingen en grondbezit. We zien dat mensen reageren op de veranderende omstandigheden door hun fysieke omgeving aan te passen en hun agrarische activiteiten te verplaatsen naar bovenstroomse gebieden waar water goedkoper en gemakkelijker te verkrijgen is, ondanks dat het illegaal gebruikt wordt. Deze casus toont ook hoe satellietbeelden gebruikt kunnen worden om complexe interacties tussen sociale en fysieke processen te doorgronden en hoe de beelden door beleidsmakers gebruikt kunnen worden om hun beleid bij te sturen.

In het laatste hoofdstuk van dit proefschrift worden de vier casussen bij elkaar gebracht in een *incorporated comparison* (McMichael, 1990; 2000). Dit is gedaan vanuit de gedachte dat de hervormingsprocessen in alle vier de landen dezelfde oorsprong hebben, en daarmee veel overeenkomsten hebben in uitgangspunten, beleidsdoelen en beleidsinstrumenten. De vergelijking laat zien dat de hervormingen in de watersectoren in deze Afrikaanse landen bijdragen aan een proces van sociaal onderscheid dat gunstig uitpakt voor historisch bevoorrechte, individueel opererende watergebruikers, die hun gewassen produceren voor de commerciële markt. We zien dat de institutionele kaders voor de waterverdeling in de bestudeerde stroomgebieden dynamisch zijn en steeds weer worden heronderhandeld, bevestigd en bevochten door de verschillende boeren. In dit proces maken boeren actief gebruik van de normatieve beginselen en institutionele modellen die door de nationale regeringen zijn geïntroduceerd als onderdeel van de hervormingsprocessen in de watersector. De boeren interpreteren, bewerken, aanvaarden en verwerpen, bewust en onbewust, onderdelen van het beleid van de overheid en combineren het met bestaande institutionele kaders tot nieuwe gemêleerde regels en bepalingen. Ook ambtenaren nemen actief deel aan dit proces en proberen de institutionele kaders zodanig te manipuleren, niet alleen om de beleidsdoelstellingen te behalen, maar ook om hun eigen inzichten en belangen te bevredigen. Overheidsbeleid wordt aldus een onderdeel van het breed institutioneel repertoire dat actoren kunnen gebruiken in de voortdurende onderhandelingsprocessen, die de institutionele kaders bepalen voor de toegang tot, de controle over en de verdeling van water. Aangezien de macht van actoren continu aan verandering onderhevig is, en daarmee ook hun vermogen om beleidsinterventies te manipuleren, concludeert dit proefschrift dat de institutionele kaders niet enkel door de hervorming van het overheidsbeleid veranderen, maar ook door ongelijke *bricolage* processen (o.a. Cleaver, 2002; 2012).

Dit onderzoek toont aan dat waterbeleid invloed heeft op de fysieke waterstromen in de onderzochte stroomgebieden, vooral wanneer het beleid is afgestemd op de belangen van de elite en geïmplementeerd wordt door middel van schijnbaar neutrale, of zelfs 'progressieve', beleidsmodellen. Beleid als zodanig kan slechts in beperkte mate bijdragen aan progressieve veranderingen in de maatschappij, vooral in het neoliberale tijdperk waarin de belangen van invloedrijke actoren binnen de nationale en internationale politiek zo met elkaar verbonden zijn en verankerd in hun normatieve beeld van de realiteit. Dit proefschrift laat de gevolgen zien van deze verschuiving naar een neoliberaal beleid waarin voornamelijk getracht wordt om institutionele processen te sturen. Technologische beleidsinstrumenten, zoals investeringen in de ontwikkeling van infrastructuur voor waterbeheer, worden nagenoeg niet ingezet. Grote delen van de waterwetgeving die is ingevoerd als onderdeel van het hervormingsproces is niet van toepassing op de meerderheid van de boeren in de bestudeerde landen, omdat zij geen toegang hebben tot (adequate) infrastructuur. In sommige gevallen leidt dit tot ongewenste resultaten, zoals verdere marginalisering van historisch benadeelde gebruikersgroepen en/of blijvende fysieke veranderingen binnen de stroomgebieden. Hiermee wordt bewezen dat gekozen beleidsinstrumenten contraproductief kunnen zijn voor het behalen van beleidsdoelstellingen. Bovendien toont dit onderzoek aan dat overheden er niet in slagen om de misstanden die voortvloeien uit het koloniale verleden aan te pakken doordat zij de middelen niet hebben om infrastructurele aanpassingen te plegen ten faveure van historisch benadeelde groepen. Die middelen kunnen ze ook niet lenen bij internationale financiële instellingen omdat dit niet past in hun neoliberale visie op waterbeheer.

Dit proefschrift draagt bij aan de bestaande theorieën en concepten met betrekking tot de institutionele veranderingsprocessen en waterbeheer, en in het bijzonder de theorie *critical institutionalism*. Dit onderzoek verrijkt deze theorie op vier manieren:

Ten eerste door in de analyse de gevolgen van de structurerende werking van institutionele veranderingsprocessen op hogere schaalniveaus mee te nemen en te bekijken hoe dat de hervormingsprocessen in stroomgebieden beïnvloedt. Om dit te bereiken is bewust gekozen voor de twee onderzoeksmethoden *extended case study method* en *incorporated comparison*. Deze helpen de wisselwerking tussen de processen op verschillende schaalniveaus te begrijpen. Daarnaast is ervoor gekozen om *critical institutionalism* te combineren met theorieën die vanuit een politiek perspectief kijken naar beleidsvormingsprocessen.

Ten tweede door te kijken naar de wisselwerking tussen sociale en fysieke processen, en in het bijzonder door te kijken naar hoe de fysieke omgeving sociale relaties tot stand brengt. Het concept van *waterscape* wordt gebruikt om de invloed van infrastructuur, alsmede de rol van de fysieke kenmerken van water te begrijpen in het vormen van de institutionele kaders in stroomgebieden.

Ten derde door het analyseren van de normatieve perspectieven waarop beleidsinterventies zijn gebaseerd en hoe deze zich verhouden tot de normen en waarden in de samenleving. Op deze manier wordt niet alleen gekeken naar hoe actoren macht uitoefenen, maar ook hoe macht verweven is in de institutionele kaders en hoe daardoor structurele ongelijkheden in de maatschappij worden geproduceerd, in stand gehouden en betwist (o.a. (e.g. Foucault, 1979, 1980).

Ten vierde door te laten zien hoe de bevindingen van dit soort studies nuttig kan zijn voor beleidsmakers. Hiervoor bevat dit proefschrift concrete suggesties voor het herzien van het huidige waterbeleid in de bestudeerde landen. Deze suggesties behelzen het erkennen en doorgronden van het politieke karakter van het beleidsvormingsproces door kritische beleidsanalyses; het tijdig reageren op onverwachte en/of ongewenste gevolgen van beleidsveranderingen; en het aannemen van alomvattend beleid dat zowel institutionele als financiële en technologische beleidsinstrumenten omvat.

Op basis van dit onderzoek worden aanbevelingen gedaan voor vervolgonderzoek. Dit behelst onder andere etnografisch onderzoek naar de actoren, die actief zijn binnen de beleidsnetwerken van waaruit de beleidsmodellen worden verspreid, alsmede onderzoek naar de invloed van de fysieke omgeving op het vormen van sociale relaties en specifiek de rol van infrastructuur hierin. Dit proefschrift sluit af met een kritisch reflectie op het onderzoek door te bespreken hoe gemaakte keuzes in de aangewende theorieën en geselecteerde methodes de uitkomsten van dit onderzoek hebben beïnvloed.

About the author

Jeltsje Sanne Kemerink-Seyoum was born in Krimpen aan den IJssel, the Netherlands, on June 8, 1979. In 2004 she obtained her MSc degree in River Engineering from the Delft University of Technology. She conducted her thesis research at WL Delft Hydraulics on a modeling assignment in which she assessed the implications of inundating the Hedwige polder on tidal flats in the Western Scheldt Estuary. During her studies she followed an internship programme with IRC for which she went to Kenya to collect field data to evaluate the technical and social aspects of the construction of sand dams using a participatory appraisal method.

After graduation Jeltsje joined UNESCO-IHE Institute for Water Education as Project Officer and became involved in the acquisition, management, monitoring and evaluation of research and capacity building projects. She maintained relations with partner organizations and donors and contributed to the successful acquisition and contract negotiation of various water related projects in different parts of the world.

In 2007 Jeltsje moved to an academic department of UNESCO-IHE and currently she holds the position of Lecturer and Researcher in Water Governance. As lecturer she delivers courses on various subjects including water governance, public participation, institutional analysis, legal pluralism and research methodology for qualitative social sciences. In 2011 she successfully obtained her University Teaching Qualification and from 2011 to 2014 she was appointed as the Programme Coordinator of the Master programme in Water Management. As Programme Coordinator she has played a key role in the successful reaccreditation of the programme. In her position as lecturer she actively supervises MSc students in their thesis research on water governance related topics and so far has guided students on case studies in amongst others Ethiopia, Indonesia, Jordan, Kenya, Rwanda, South Africa, Sudan, Vietnam and Zimbabwe. Early 2015 Jeltsje was appointed as an affiliated faculty member of the Oregon State University in the United States of America.

In 2007 Jeltsje started her PhD thesis research on a part time basis. Her research focuses on the implications of water reform processes on access to and control over water resources for various agricultural water users in catchments in Kenya, Tanzania, South-Africa and Zimbabwe. In addition to her PhD research, she is involved in other interdisciplinary research on socio-nature processes in Asia and Africa. She contributed to research in Bangladesh that studies the response of society on flood as well as the distribution of flood risk in society, and she participates in research on the implications of climate financing mechanisms on access to land and water resources in Ethiopia and Indonesia.

Besides her research activities, Jeltsje has been involved in various capacity development projects in the field of water and environmental management in amongst other Rwanda, South Africa, Vietnam and Zimbabwe. In addition, she has been involved in an advisory project on climate change adaptation for coastal villages on the islands of São Tomé and Principe for which she organized and facilitated the stakeholder consultation meetings.

List of publications

Di Baldassarre, G., **J. S. Kemerink**, M. Kooy, L. Brandimarte (2014) Floods and Societies: the Spatial Distribution of Water-related Disaster Risk and its Dynamics. *WIREs Water* 1(2):133-139.

Kemerink, J.S., L. E. Méndez, R. Ahlers, P. van der Zaag (2013) The question of inclusion and representation in rural South Africa: challenging the concept of Water User Associations as a vehicle for transformation. *Water Policy* 15: 243–257.

Di Baldassarre, G., M. Kooy, **J. S. Kemerink**, L. Brandimarte (2013) Towards understanding the dynamic behaviour of floodplains as human-water systems. *Hydrol. Earth Syst. Sci.* 10: 3869–3895.

Kemerink, J.S., D. Mbuvi, Schwartz, K. (2012) Governance shifts in the water services sector: a case study of the Zambia water services sector. In Katko T., Juuti P.S. and Schwartz K. (eds) *Water Services Management and Governance: Lessons for a Sustainable Future.* IWA Publishing: 3-11.

Komakech, H.C., P. van der Zaag, M. L. Mul, T.A. Mwakalukwa, **J.S. Kemerink** (2012) *Formalization of water allocation systems and impacts on local practices in the Hingilili subcatchment, Tanzania.* International Journal of River Basin Management, 10 (3): 213-227.

Kemerink, J.S., R. Ahlers, P. van der Zaag (2011) Contested water right in post-apartheid South-Africa: the struggle for water at catchment level. *Water SA*, 37 (4): 585-594.

Mul, M.L., **J.S. Kemerink**, N.F. Vyagusa, M.G. Mshana, P. van der Zaag, and H. Makurira (2011) Water allocation practices among smallholder farmers in the South Pare Mountains, Tanzania; can they be up-scaled? *Agricultural Water Management* 98(11): 1752-1760.

Kemerink, J.S., R. Ahlers, P. van der Zaag (2009) Assessment of the potential for hydro-solidarity in plural legal condition of traditional irrigation systems in northern Tanzania. *Physics and Chemistry of the Earth* 34: 881-889.

Kemerink, J.S., S.N. Munyao, K. Schwartz, R. Ahlers, P. van der Zaag (*forthcoming*) Why infrastructure still matters: unravelling water reform processes in an uneven waterscape in rural Kenya. Under review *International Journal of the Commons.*

Kemerink, J.S., N.L. T. Chinguno, S.D. Seyoum, R. Ahlers, P. van der Zaag (*forthcoming*) Jumping the water queue: changing waterscapes under water reform processes in rural Zimbabwe. Under review *Natural Resources Forum.*

Méndez, L. E., **J.S. Kemerink**, P. Wester, F. Molle (*forthcoming*) The quest for water: Strategizing water control and circumventing reform in rural South Africa. To be submitted to *International Journal on Water Resource Development.*